Information and Instructions

This shop manual contains several sections each covering a specific group of wheel type tractors. The Tab Index on the preceding page can be used to locate the section pertaining to each group of tractors. Each section contains the necessary specifications and the brief but terse procedural data needed by a mechanic when repairing a tractor on which he has had no previous actual experience.

Within each section, the material is arranged in a systematic order beginning with an index which is followed immediately by a Table of Condensed Service Specifications. These specifications include dimensions, fits, clearances and timing instructions. Next in order of arrangement is the procedures paragraphs.

In the procedures paragraphs, the order of presentation starts with the front axle system and steering and proceeding toward the rear axle. The last paragraphs are devoted to the power take-off and power lift systems. Interspersed where needed are additional tabular specifications pertaining to wear limits, torquing, etc.

HOW TO USE THE INDEX

Suppose you want to know the procedure for R&R (remove and reinstall) of the engine camshaft. Your first step is to look in the index under the main heading of ENGINE until you find the entry "Camshaft." Now read to the right where under the column covering the tractor you are repairing, you will find a number which indicates the beginning paragraph pertaining to the camshaft. To locate this wanted paragraph in the manual, turn the pages until the running index appearing on the top outside corner of each page contains the number you are seeking. In this paragraph you will find the information concerning the removal of the camshaft.

More information available at haynes.com
Phone: 805-498-6703

J H Haynes & Co. Ltd.
Haynes North America, Inc.

ISBN-10: 0-87288-107-5
ISBN-13: 978-0-87288-107-5

Cover art by Sean Keenan

Disclaimer

SHOP MANUAL
INTERNATIONAL HARVESTER

SERIES

706	1206	2706	2856
756	1256	2756	21206
806	1456	2806	21256
856			21456

Engine serial number is stamped on left side of engine crankcase on all series except series 706, 2706, 756 and 2756 equipped with the D-310 diesel engine. Engine serial number is stamped on right side of engine crankcase on the D-310 engine. Engine serial number will be preceded by engine model number.

Tractor serial number is stamped on name plate attached to right side of clutch housing.

INDEX (By Starting Paragraph)

CONDENSED SERVICE DATA

	Early Series 706-2706		Late Series 706-2706, 756-2756		Series 806-2806
	Diesel	Non-Diesel	Diesel	Non-Diesel	Diesel

GENERAL

Engine Make	IH	IH	IH	IH	IH
Engine Model	D282	C263	D310	C291	D361
Number of Cylinders	6	6	6	6	6
Bore—Inches	3 11/16	3 9/16	3⅞	3¾	4⅛
Stroke—Inches	4 25/64	4 25/64	4⅜	4 25/64	4½
Displacement—Cubic Inches	282	263	310	291	361
Pistons Removed From	Top	Top	Top	Top	Top
Main Bearings, Number of	4	4	7	4	7
Main and Rod Bearings, Adjustable?	No	No	No	No	No
Cylinder Sleeves	Dry	Dry	Wet	Dry	Dry
Forward Speeds Without T.A.	8	8	8	8	8
Forward Speeds With T.A.	16	16	16	16	16
Alternator, Generator and Starter Make	Delco-Remy	Delco-Remy	Delco-Remy	Delco-Remy	Delco-Remy

TUNE-UP

Compression Pressure Except LP-Gas	350-400 (1)	175 (1)	315-340 (1)	190 (1)	365-410 (1)
LP-Gas	......	205 (1)	......	205 (1)	......
Firing Order	1-5-3-6-2-4	1-5-3-6-2-4	1-5-3-6-2-4	1-5-3-6-2-4	1-5-3-6-2-4
Valve Tappet Gap (Hot) Intake	.027	.027	.010	.027	.013
Valve Tappet Gap (Hot) Exhaust	.027	.027	.012	.027	.025
Inlet Valve Seat Angle (Degrees)	45°	30°	45°	30°	45°
Exhaust Valve Seat Angle (Degrees)	45°	30°	45°	30°	45°
Ignition Distributor Make	......	IH	......	IH (2)	......
Breaker Gap	......	.020	......	.020	......
Distributor Timing, Retard (Gasoline)	......	TDC	......	1° BTDC	......
Distributor Timing, Advanced (Gasoline)	......	22° BTDC	......	18° BTDC	......
Distributor Timing, Retard (LP-Gas)	......	2° BTDC	......	2° BTDC	......
Distributor Timing, Advanced (LP-Gas)	......	24° BTDC	......	24° BTDC	......
Timing Mark Location	Crankshaft Pulley or Flywheel				Flywheel
Spark Plug Electrode Gap Gasoline	......	.023	......	.023	......
LP-Gas	......	.015	......	.015	......
Carburetor Make, LP-Gas	......	Ensign	......	Ensign	......
Carburetor Make, Gasoline	......	IH	......	IH	......
Battery Terminal, Grounded	Negative	Negative	Negative	Negative	Negative
Engine Low Idle rpm	650	425	650	425	650
Engine High Idle rpm, No Load	2507	2530	2530	2530	2620
Engine Full Load rpm	2300	2300	2300	2300	2400

(1) Approximate psi, at sea level, at cranking speed.
(2) Delco-Remy, if equipped with Magnetic Pulse ignition.

SIZES—CAPACITIES—CLEARANCES
(Clearance in Inches)

Crankshaft Main Journal Diameter	2.7480-2.7490	2.7480-2.7490	3.1484-3.1492	2.7480-2.7490	3.3742-3.3755
Crankpin Diameter	2.3730-2.3740	2.3730-2.3740	2.5185-2.5193	2.3730-2.3740	2.9980-2.9990
Camshaft Journal Diameter, No. 1 (Front)	2.1090-2.1100	2.1090-2.1100	*	2.1090-2.1100	2.4290-2.4300
Camshaft Journal Diameter, No. 2	2.0890-2.0900	2.0890-2.0900	*	2.0890-2.0900	2.0890-2.0900
Camshaft Journal Diameter, No. 3	2.0690-2.0700	2.0690-2.0700	*	2.0690-2.0700	2.0690-2.0700
Camshaft Journal Diameter, No. 4	1.4990-1.5000	1.4995-1.5005	*	1.4995-1.5005	1.4990-1.5000
Piston Pin Diameter, Diesel	1.1247-1.1249	......	1.4172-1.4173	......	1.4998-1.5000
Non-Diesel	......	0.8748-0.8749	......	0.8748-0.8749	......
Valve Stem Diameter					
Inlet	0.3715-0.3725	0.3715-0.3725	0.3919-0.3923	0.3715-0.3725	0.4348-0.4355
Exhaust	0.3715-0.3725	0.3710-0.3720	0.3911-0.3915	0.3710-0.3720	0.4348-0.4355
Main Bearing Diametral Clearance	0.0012-0.0042	0.0012-0.0042	0.0029-0.0055	0.0012-0.0042	0.0018-0.0051
Rod Bearing Diametral Clearance	0.0009-0.0034	0.0009-0.0034	0.0023-0.0048	0.0009-0.0034	0.0018-0.0051
Piston Skirt Diametral Clearance	0.0050-0.0076	0.0010-0.0045	0.0039-0.0047	0.0010-0.0045	0.0038-0.0056
Crankshaft End Play	0.0050-0.0130	0.0050-0.0130	0.0060-0.0090	0.0050-0.0130	0.0070-0.0185
Camshaft Bearings Diametral Clearance	0.0005-0.0050	0.0005-0.0050	0.0009-0.0033	0.0005-0.0050	0.0010-0.0055
Camshaft End Play	0.0020-0.0100	0.0020-0.0100	0.0040-0.0170	0.0020-0.0100	0.0020-0.0100
Cooling System Capacity—Quarts	21½	20½	23½	20½	24
Crankcase Oil, Quarts	9	9	12	9	13
Rear Axle Housing (Each), Quarts	6	6	6	6	5
Transmission and Differential—					
Gallons (approximate)	17	17	17	17	17
Front Differential Housing					
(All-wheel drive tractor), Quarts	10	10	10	10	10

*All seven journals are 2.2823-2.2835.

TIGHTENING TORQUES—FOOT POUNDS

Camshaft Nut	110-120	110-120	......	110-120	50-60
Rod Bearing Screws	See Par. 77 & 133		63	See Par. 77	105
Cylinder Head Screws	110-120	85-85	90	85-95	135
Flywheel Screws	60	75	100	75	95
Injection Nozzle Hold Down Screws	20-25	......	8-9	......	10-11
Injection Nozzle Holder Nut	45-50	......	50	......	65
Main Bearing Screws	80	80	See Par. 135	80	115
Manifold Screws	20-25	45	37	45	55

CONDENSED SERVICE DATA

	Series 806-2806, 856-2856 Non-Diesel	Series 856-2856 Diesel	Series 1206-21206 Turbo-Diesel	Series 1256-21256 Turbo-Diesel	Series 1456-21456 Turbo-Diesel
GENERAL					
Engine Make	IH	IH	IH	IH	IH
Engine Model	C301	D407	DT361	DT407	DT407
Number of Cylinders	6	6	6	6	6
Bore—Inches	3 13/16	$4\frac{5}{16}$	$4\frac{1}{8}$	$4\frac{5}{16}$	$4\frac{5}{16}$
Stroke—Inches	4 25/64	$4\frac{5}{8}$	$4\frac{1}{2}$	$4\frac{5}{8}$	$4\frac{5}{8}$
Displacement—Cubic Inches	301	407	361	407	407
Pistons Removed From	Top	Top	Top	Top	Top
Main Bearings, Number of	4	7	7	7	7
Main and Rod Bearings, Adjustable?	No	No	No	No	No
Cylinder Sleeves	None	Dry	Dry	Dry	Dry
Forward Speeds Without T.A.	8	8	8	8	8
Forward Speeds With T.A.	16	16	16	16	16
Alternator, Generator and Starter Make	Delco-Remy	Delco-Remy	Delco-Remy	Delco-Remy	Delco-Remy
TUNE-UP					
Compression Pressure Except LP-Gas	185 (1)	350-425 (1)	350-410 (1)	350-425 (1)	350-425 (1)
LP-Gas	210 (1)	......	......	......	......
Firing Order	1-5-3-6-2-4	1-5-3-6-2-4	1-5-3-6-2-4	1-5-3-6-2-4	1-5-3-6-2-4
Valve Tappet Gap (Hot) Intake	.027	.013	.013	.013	.013
Valve Tappet Gap (Hot) Exhaust	.027	.025	.025	.025	.025
Inlet Valve Seat Angle (Degrees)	30°	45°	45°	45°	45°
Exhaust Valve Seat Angle (Degrees)	30°	45°	45°	45°	45°
Ignition Distributor Make	IH (2)	......	......	......	......
Breaker Gap	.020	......	......	......	......
Distributor Timing, Retard (Gasoline)	TDC	......	......	......	......
Distributor Timing, Advanced (Gasoline)	22° BTDC	......	......	......	......
Distributor Timing, Retard (LP-Gas)	2° BTDC	......	......	......	......
Distributor Timing, Advanced (LP-Gas)	24° BTDC	......	......	......	......
Timing Mark Location	Flywheel	Flywheel	Flywheel	Flywheel	Flywheel
Spark Plug Electrode Gap Gasoline	.023	......	......	......	......
LP-Gas	.015	......	......	......	......
Carburetor Make, LP-Gas	Ensign	......	......	......	......
Carburetor Make, Gasoline	IH	......	......	......	......
Battery Terminal, Grounded	Negative	Negative	Negative	Negative	Negative
Engine Low Idle rpm	425	650	675	650	650
Engine High Idle rpm, No Load	2640	2650	2630	2650	2650
Engine Full Load rpm	2400	2400	2400	2400	2400

(1) Approximate psi, at sea level, at cranking speed.

(2) Delco-Remy, if equipped with Magnetic Pulse ignition.

SIZES—CAPACITIES—CLEARANCES					
(Clearance in Inches)					
Crankshaft Main Journal Diameter	2.7480-2.7490	3.3742-3.3755	3.3742-3.3755	3.3742-3.3755	3.3742-3.3755
Crankpin Diameter	2.3730-2.3740	2.9980-2.9990	2.9980-2.9990	2.9980-2.9990	2.9980-2.9990
Camshaft Journal Diameter, No. 1 (Front)	2.1090-2.1100	2.4290-2.4300	2.4290-2.4300	2.4290-2.4300	2.4290-2.4300
Camshaft Journal Diameter, No. 2	2.0890-2.0900	2.0890-2.0900	2.0890-2.0900	2.0890-2.0900	2.0890-2.0900
Camshaft Journal Diameter, No. 3	2.0690-2.0700	2.0690-2.0700	2.0690-2.0700	2.0690-2.0700	2.0690-2.0700
Camshaft Journal Diameter, No. 4	1.4995-1.5005	1.4990-1.5000	1.4990-1.5000	1.4990-1.5000	1.4990-1.5000
Piston Pin Diameter, Diesel	......	1.4998-1.5000	1.4998-1.5000	1.4998-1.5000	1.6248-1.6250
Non-Diesel	0.8748-0.8749	......	......	......	......
Valve Stem Diameter					
Inlet	0.3715-0.3725	0.4348-0.4355	0.4348-0.4355	0.4348-0.4355	0.4348-0.4355
Exhaust	0.3710-0.3720	0.4348-0.4355	0.4348-0.4355	0.4348-0.4355	0.4348-0.4355
Main Bearing Diametral Clearance	0.0012-0.0042	0.0018-0.0051	0.0018-0.0051	0.0018-0.0051	0.0018-0.0051
Rod Bearing Diametral Clearance	0.0009-0.0034	0.0018-0.0051	0.0018-0.0051	0.0018-0.0051	0.0018-0.0051
Piston Skirt Diametral Clearance	0.0010-0.0045	0.0049-0.0069	0.0038-0.0056	0.0049-0.0069	0.0049-0.0069
Crankshaft End Play	0.0050-0.0130	0.0070-0.0185	0.0070-0.0185	0.0070-0.0185	0.0070-0.0185
Camshaft Bearings Diametral Clearance	0.0005-0.0050	0.0010-0.0055	0.0010-0.0055	0.0010-0.0055	0.0010-0.0055
Camshaft End Play	0.0020-0.0100	0.0020-0.0100	0.0020-0.0100	0.0020-0.0100	0.0020-0.0100
Cooling System Capacity—Quarts	21	24	23½	23½	23½
Crankcase Oil, Quarts	9	11	15	13	13
Rear Axle Housing (Each), Quarts	5	5	5	5	5
Transmission and Differential—					
Gallons (approximate)	17	17	17	17	22
Front Differential Housing (All-wheel drive tractors), Quarts	10	10	10	10	10

TIGHTENING TORQUES—FOOT POUNDS					
Camshaft Nut	110-120	55	55	55	55
Rod Bearing Screws	See Par. 77	105	105	105	105
Cylinder Head Screws	85-95	135	135	135	135
Flywheel Screws	75	95	95	95	95
Injection Nozzle Hold Down Screws	......	10-11	10-11	10-11	10-11
Injection Nozzle Holder Nut	......	65	65	65	65
Main Bearing Screws	80	115	115	115	115
Manifold Screws	45	55	55	55	55

FRONT SYSTEM TRICYCLE TYPE

Farmall tractors are available with either of the two single front wheels shown in Fig. IH100, or may also be equipped with a dual wheel tricycle type shown in Fig. IH101.

SINGLE WHEEL

Farmall Models

1. The single front wheel is mounted in a fork which is bolted to the steering pivot shaft and depending on the tire size, two wheel types are used. On tractors with 9.00 x 10 or 11.00 x 10 tires, the male and female wheel halves (18 and 19—Fig. IH100) are used. On tractors with 7.50 x 16 or 7.50 x 20 tires a solid disc type wheel (12) is used in conjunction with hub (11). In all cases taper roller wheel bearings are used.

Wheel bearings are adjusted to a slight preload with adjusting nut (14). Lock adjusting nut with jam nut (16) after adjustment is complete.

DUAL WHEELS

Farmall Models

2. The pedestal for the dual tricycle wheels is bolted to the steering pivot shaft. The pedestal is available as a pre-riveted assembly (17—Fig. IH101), or the pedestal and axle (18) are available as separate repair parts. Wheel bearings are adjusted to a slight preload.

FRONT SYSTEM AXLE TYPE

AXLE MAIN MEMBER

Farmall Models

3. For Farmall tractors equipped with an adjustable wide tread front axle, refer to Fig. IH102 or IH103. The axle main member (24) pivots on pin (26) which is pinned in the axle support (9). The two pivot pin bushings (25) are pressed into the axle main member and should be reamed after installation, if necessary, to provide a free fit for the pivot pin.

To remove the axle main member, disconnect tie rods from steering arms, then remove axle clamps (21) and withdraw the axle extension, knuckle and wheel assemblies. Remove cotter pin and pivot pin retaining pin, then

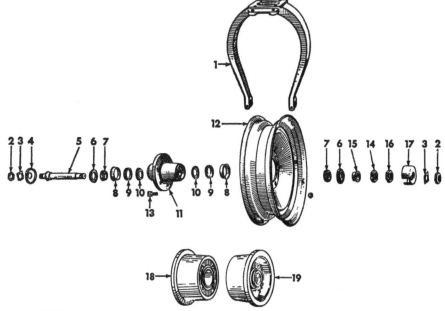

Fig. IH100—Exploded view showing both types of fork mounted single front wheels. Axle (5) is fitted with taper roller bearings and is used with both types of wheels.

1. Fork	8. Seal retainer	15. Spacer
2. Jam nut	9. Bearing cup	16. Jam nut
3. Lock washer	10. Grease retainer	17. Shield (long)
4. Shield (short)	11. Hub	18. Wheel half
5. Axle	12. Wheel	(male)
6. Oil seal	13. Wheel bolt	19. Wheel half
7. Bearing cone	14. Adjusting nut	(female)

disconnect stay rod ball from its support. Save the shims (5) located between socket (4) and socket cap (6). Drive pivot pin out of axle support and axle main member and remove

axle main member. On all models except Farmall 1456, stay rod (8) can now be removed from axle main member if necessary. The stay rod and axle main member (24—Fig. IH103)

Fig. IH101 — Exploded view of front support, tricycle type front axle and components used on Farmall tractors.

1. Front frame R.H.
2. Front frame L.H.
3. Grease fitting
5. Cover
6. Shaft nut
7. Bearing cone
8. Bearing cup
9. "O" ring
10. Plug
11. Front support
12. "O" ring
13. Bearing cup
14. Bearing cone
15. Oil seal
16. Steering pivot shaft
17. Pedestal
18. Axle
19. Dust shield
20. Pulley shield

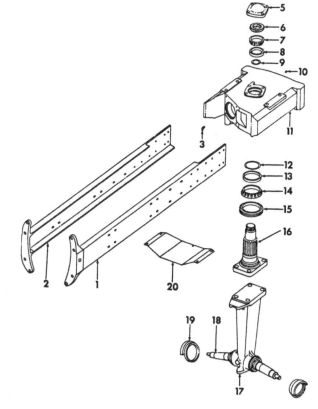

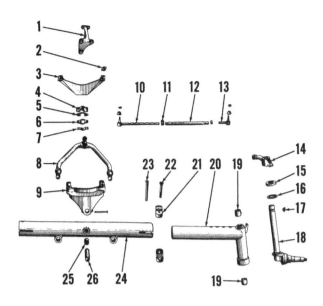

Fig. IH102 — Adjustable wide front axle available for all Farmall tractors except 1456. Refer also to Figs. IH103 and IH-104.

1. Steering arm (center)
2. Washer (4 used)
3. Stay rod support
4. Ball socket
5. Shim
6. Socket cap
7. Lock plate
8. Stay rod assembly
9. Axle support
10. Tie rod extension
11. Clamp
12. Tube
13. Tie rod end
14. Steering arm
15. Thrust bearing
16. Felt washer
17. Woodruff key
18. Steering knuckle
19. Bushing
20. Axle extension
21. Axle clamp
22. Clamp bolt
23. Clamp pin
24. Axle main member
25. Pivot bushing
26. Pivot pin

is a welded assembly on Farmall 1456 tractors.

Reinstall by reversing the removal procedure and if necessary, adjust the front wheel toe-in as outlined in paragraph 9.

Farmall 706-756-806-856
Hi-Clear

4. Farmall 706, 756, 806 and 856 high clearance tractors are equipped with a front axle such as that shown in Fig. IH104.

Removal procedure for this axle is the same as given in paragraph 3 except that the auxiliary stay rods (33) should be disconnected when removing the axle extension and wheel assemblies. Refer to paragraph 9 for toe-in adjustment.

1. Steering arm (center)
2. Washer (4 used)
3. Stay rod support
4. Ball socket
5. Shim
6. Socket cap
7. Lock plate
9. Axle support
10. Tie rod extension
11. Clamp
12. Tube
13. Tie rod end
14. Steering arm
15. Thrust bearing
16. Felt washer
18. Steering knuckle
19. Bushing
20. Axle extension
21. Axle clamp
24. Axle main member
25. Pivot bushing
26. Pivot pin
34. Tie rod assembly
35. Retaining ring

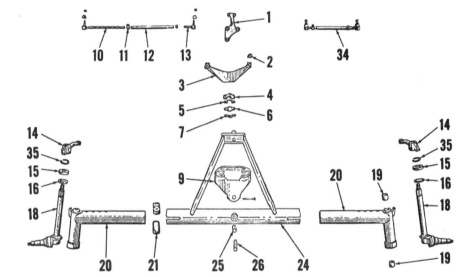

Fig. IH103—Adjustable wide front axle available for Farmall 1456 tractors.

1. Steering arm (center)
4. Ball socket
5. Shim
6. Socket cap
7. Lock plate
8. Stay rod
9. Axle support
10. Tie rod extension
11. Clamp
12. Tube
13. Tie rod end
14. Steering arm
15. Thrust bearing
16. Felt washer
17. Woodruff key
18. Steering knuckle
19. Bushing
20. Axle extension
21. Axle clamp
22. Clamp bolt
23. Clamp pin
24. Axle main member
25. Pivot bushing
26. Pivot pin
27. Adjustable end
28. Clamp
29. Tube
30. Threaded end
31. Upper bracket
32. Lower bracket
33. Auxiliary stay rod assembly

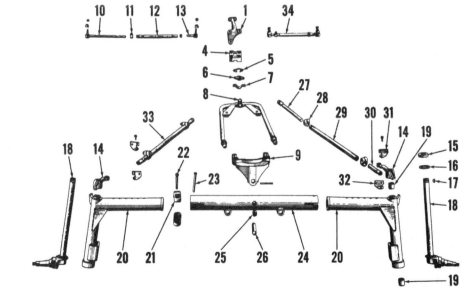

Fig. IH104—Exploded view of adjustable wide front axle used on Farmall "Hi-Clear" models. Note the auxiliary stay rod assemblies.

International Models

5. International tractors are available with a non-adjustable as well as an adjustable front axle as shown in Figs. IH105 and IH106. Stay rod, stay rod ball and axle main member of both axles are available separately for service. The axle, or axle main member, pivots on pin (14—Fig. IH105 or IH106) which is retained in the axle support with a pin and a bolt and nut. The two pivot pin bushings are pressed into the axle main member, and should be reamed after installation, if necessary, to provide a free fit for the pivot pin.

6. To remove the non-adjustable axle proceed as follows: Remove nut and washer from top of steering knuckles, place correlation marks on steering arms and knuckles and with tie rods attached, remove steering arms from knuckles. Remove retaining rings (7—Fig. IH105), then raise front of tractor and withdraw steering knuckles and wheels as assemblies. Remove pivot pin retaining bolt and pin from axle support, disconnect stay rod ball and save the shims (20) located between socket (19) and cap (21). Axle main member and stay rod can now be removed after driving out pivot pin. Stay rod can be separated from axle, if necessary.

Reinstall by reversing removal procedure and refer to paragraph 9 if necessary to adjust toe-in.

7. To remove the adjustable axle main member (9—Fig. IH106), proceed as follows: Remove nut and washer from top of steering arms, place correlation marks on steering arms and knuckles, then with tie rods attached, remove steering arms from knuckles. Remove axle clamps (24), raise front of tractor and remove the axle extensions and wheels as assemblies. Remove pivot pin retaining bolt and pin from axle support, disconnect stay rod ball and save shims (20) located between socket (19) and cap (21). Axle main member and stay rod (18) can be removed after driving out pivot pin (14) Stay rod can be separated from axle main member if necessary.

Reinstall by reversing removal procedure and refer to paragraph 9 if necessary to adjust toe-in.

STAY ROD AND BALL

All Models

8. The stay rod and stay rod ball in all cases are available as individual parts. Clearance between stay rod ball and socket can be adjusted by adding or subtracting shims which are located between socket and cap. The socket cap retaining cap screws are tightened to 85-100 ft.-lbs. torque and locked by bending tabs of lock plate against flats of screw heads. On all models except Farmall 1456, the stay rod is detachable from axle or axle main member. Farmall models equipped with adjustable wide tread front axle have a socket support bolted to side rails and any service required on support is obvious.

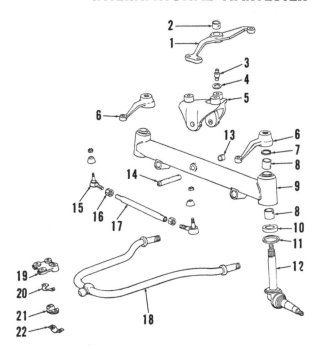

Fig. IH105 — Non-adjustable front axle and component parts available for all International tractors. Refer also to Fig. IH106.

1. Steering arm (center)
2. Bushing
3. Pivot pin
4. Shim (0.010 and 0.020)
5. Axle support
6. Steering arm
7. Retaining ring
8. Bushing
9. Axle main member
10. Thrust bearing
11. Felt washer
12. Steering knuckle
13. Pivot bushing
14. Pivot pin
15. Tie rod end
16. Jam nut
17. Tube
18. Stay rod
19. Ball socket
20. Shim
21. Socket cap
22. Lock plate

TIE RODS AND TOE-IN

All Models

9. The procedure for removal and disassembly of the tie rods on all models so equipped is obvious after an examination of the units. Tie rod ends are non-adjustable and faulty units will require renewal.

Adjust the toe-in on all models to ¼-inch, plus or minus $\frac{1}{16}$-inch. Adjustment is made by varying the length of

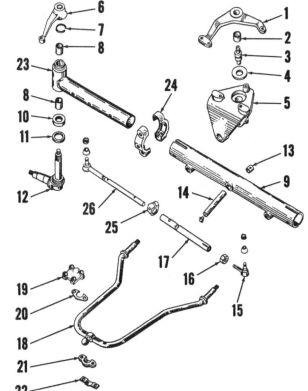

Fig. IH106 — Adjustable front axle and component parts available for all International tractors.

1. Steering arm (center)
2. Bushing
3. Pivot pin
4. Shim (0.010 and 0.020)
5. Axle support
6. Steering arm
7. Retaining ring
8. Bushing
9. Axle main member
10. Thrust bearing
11. Felt washer
12. Steering knuckle
13. Pivot bushing
14. Pivot pin
15. Tie rod end
16. Jam nut
17. Tube
18. Stay rod
19. Ball socket
20. Shim (0.010 & 0.020)
21. Socket cap
22. Lock plate
23. Axle extension
24. Axle clamp

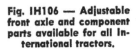

the tie rods. Both tie rods should be adjusted an equal amount with not more than one turn difference when adjustment is complete.

STEERING KNUCKLES

All Models

10. Removal of steering knuckles from axle extensions, or axle main member is obvious after an examination of the unit and reference to Figs. IH102, IH103, IH104, IH105 and IH-106. Note that the steering knuckles used on International model tractors (Figs. IH105 and IH106) and Farmall 1456 (Fig. IH103) have a retaining ring located at top end of knuckle pivot shaft.

When renewing steering knuckle bushings be sure to align oil hole in bushing with oil hole in axle or axle extension. Install bushings so outer ends are flush with bore. Ream the bushings after installation, if necessary, to provide an operating clearance of 0.001-0.006.

FRONT SYSTEM ALL-WHEEL DRIVE

11. Farmall and International tractors are available as 4-wheel drive (All Wheel Drive) units. The front axle assembly consists of a one-piece center housing having flanged ends to which stub axle ends are bolted. Wheel spindles and wheel hubs are carried on taper roller bearings. The axle center housing incorporates a straddle mounted pinion and a four pinion differential gear unit. Full floating axles extend outward from the differential and attach to Cardan type universal joints located inside wheel hubs.

Power for the front axle assembly is taken from a gear reduction unit (transfer case) which is bolted to the left side of tractor rear frame and driven by an idler gear on the reverse idler shaft of the range transmission. The reduction unit and the front axle are connected by a drive shaft fitted with two conventional universal joints of which the rear has a slip joint that compensates for oscillation of the front axle.

The gear reduction unit (transfer case) has a shifting mechanism which permits shifting to neutral, thus disconnecting power to the front axle. Gear sets with various ratios are available for the reduction unit to match the front and rear tire size combinations. Therefore, it is essential that the

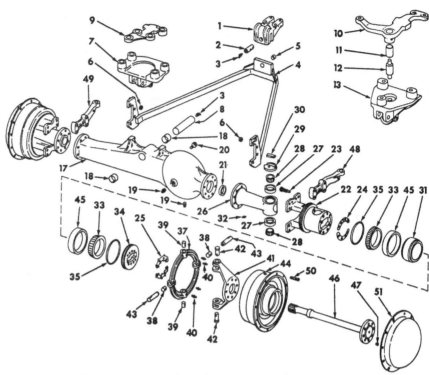

Fig. IH107—Exploded view of the front drive axle used when tractors are equipped with "All Wheel Drive". Note items 10 through 13 which are used on International tractors and differ from items 7 and 9 which are used on Farmall tractors.

1. Stay bar support	17. Housing	33. Wheel bearing
2. Rear pivot pin	18. Bushing	34. Bearing plate
3. Grease fitting	19. Plug	35. Seal
4. Stay rod	20. Vent	37. Compensating
5. Bushing	21. Oil seal	ring
6. Gusset washer	22. Spindle	38. Bushing
7. Axle support	23. Wedge adjusting	39. Bushing
(Farmall)	screw	40. Grease fitting
8. Pivot pin	24. Shims	41. Power yoke
9. Center steering	25. Lock plates	42. Yoke pin
arm (Farmall)	26. Axle stub	43. Hub pin
10. Center steering	27. Bearing cup	45. Bearing cup
arm	28. Pivot bearing	46. Axle shaft
(International)	29. Bearing cap	47. Tapered bushing
11. Bushing	30. Adjusting wedge	48. Steering arm, LH
12. Pivot pin	31. Clamp ring	49. Steering arm, RH
13. Axle support	32. Grease fitting	50. Wheel stud
(International)		51. Wheel cover

correct front and rear tire sizes be used with the correct gear set. If a change in tire sizes or gear sets is contemplated, contact the International Harvester Company for information concerning the proper combinations.

WHEEL AND PIVOT BEARINGS

Models So Equipped

12. **WHEEL BEARINGS.** Wheel bearings should be removed, cleaned and repacked annually. Removal of inner wheel bearing requires removal of hub assembly.

To remove both wheel bearings, refer to Fig. IH107 and proceed as follows: Support axle assembly, then remove wheel cover (51) and the wheel and tire assembly. Clip and remove lock wire from axle retaining cap screws, then remove cap screws and tapered bushings (47) and remove axle by pulling straight outward as shown in Fig. IH108. NOTE: Use cau-

tion when removing axle shaft not to damage the oil seal (21—Fig. IH107) located in center housing as shown in Fig. IH109. Swing power yoke (41—Fig. IH108) aside, straighten tabs of

Fig. IH108—View showing axle (46) being removed. Item (41) is power yoke. Refer also to Fig. IH109.

Fig. IH109 — When removing axle use caution not to damage the oil seal (21) shown. Differential and carrier have been removed for clarity.

retainer plates (25—Fig. IH110), then unbolt and remove bearing plate (34) and shims (24). Outer wheel bearing, power yoke, compensating ring and hub can now be removed from spindle. See Fig. IH111.

To remove inner wheel bearing (33—Fig. IH107) from spindle (22), loosen jam nut and adjusting screw (23), then position a small punch against outer end of adjusting wedge (30—Fig. IH111) and bump wedge inward until clamp ring (31) is free and remove clamp ring. Insert a pin punch through knock-out holes provided in the spindle and bump on inner side of bearing inner race until bearing is about ½-inch from inner flange of spindle, then attach puller, if necessary, and complete removal of bearing. Be sure to keep bearing straight while removing or damage (scoring) to spindle could result.

NOTE: In some cases, the top pivot bearing cap may come out of bore in spindle when clamp ring and inner wheel bearing are removed and the spindle may drop and rest on the stub axle. If this occurs, proceed as fol-

lows when reassembling. Place a wood block under spindle, then with a hydraulic jack under block, raise spindle into its proper position and place upper pivot bearing and cap in its bore. Start inner wheel bearing on spindle and over upper bearing cap. This will hold parts in position. Start clamp ring over spindle and complete installation of both bearing and clamp ring.

13. Clean and inspect all parts for wear, excessive scoring or other damage and renew as necessary. Repack the wheel bearings with a good grade of multi-purpose lithium grease.

14. Reassemble and adjust wheel bearings and pivot bearings as follows: Install inner wheel bearing on spindle with largest diameter toward inside. Place clamp ring over spindle and against inner wheel bearing and make sure it does not contact the cage of inner wheel bearing. Tighten adjusting wedge until it feels solid, then measure distance from end of wedge to end of slot in pivot bearing cap. This distance (D—Fig. IH113) should be at least 1⅜ inches. If measured distance is less than 1⅜ inches, remove spindle and add shims under bottom pivot bearing cone. Shims can be made from shim stock and each 0.012 shim will change the measured distance about ⅛-inch.

NOTE: This operation is to insure that a satisfactory adjustment can be made for the pivot bearings.

With clamp ring installed as outlined above and a new seal on spindle, install the hub, compensating ring, power yoke assembly and outer wheel bearing. With new seal on bearing plate, install original shims, bearing plate and lock plates. Tighten cap screws securely and check rotation of wheel hub. Hub should rotate with a slight drag. Add or subtract shims under bearing plate to obtain correct bearing adjustment. Shims are available in thicknesses of 0.002, 0.010 and 0.030. Lock cap screws with lock plates when adjustment is complete.

Disconnect tie rod from steering arm, then tighten wedge adjusting screw until a slight drag is felt when moving spindle through its full range of travel. Move spindle back and forth several times, then using a spring scale attached to tie rod hole of steering arm, check the pounds pull required to keep spindle in motion. See Fig. IH114. Pull should not exceed 12 pounds and must be read while spindle is in motion. Tighten adjusting screw jam nut when adjustment is complete. Reinstall axle using caution not to damage oil seal located in cen-

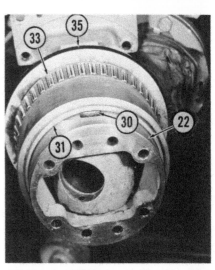

Fig. IH111—Spindle assembly with outer wheel bearing and hub removed. Note outer end of pivot bearing adjusting wedge (30).

22. Spindle	33. Inner bearing
30. Adjusting wedge	35. Seal
31. Clamp ring	

ter housing. Install tapered bushings and the axle retaining cap screws and tighten cap screws securely. Lock wire the axle cap screws. Reinstall wheel and tire and wheel cover and mate holes of wheel cover with grease fittings of compensating ring if tractor is equipped with this type wheel cover.

15. **PIVOT BEARINGS.** To remove the pivot bearings, the wheel bearings must be removed as outlined in paragraph 12.

With wheel bearings removed, disconnect tie rod from steering arm, then remove pivot bearing cups from their recesses in the stub axle by driv-

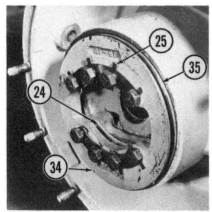

Fig. IH110—View showing bearing adjusting plate and shims. Power yoke and compensating ring are removed for clarity.

24. Shims	34. Bearing plate
25. Lock plates	35. Seal

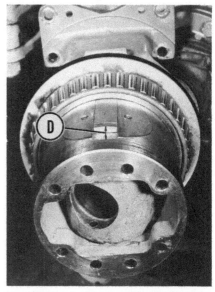

Fig. IH113—A distance (D) of 1⅜ inches between end of adjusting wedge and end of slot in pivot bearing cap should be maintained. Refer to text.

Fig. IH114 — Attach spring scale as shown to check preload of pivot bearings. Refer to text.

Fig. IH117—Support spindle on a wood block during installation of pivot bearings.

ing on a punch inserted in the knock-out holes (H—Fig. IH115) provided in outer end of stub axle. Use caution during this operation not to let punch slip past cups and damage bearings. With bearing cups free of their recesses, tilt inner end of spindle upward and remove from stub axle. Do not use force to remove spindle. If spindle does not come off readily, it is probable that bearing cups are not completely free of their recesses. The

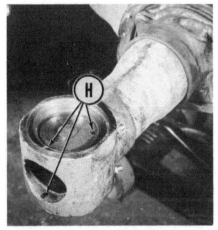

Fig. IH115—Knock-out holes (H) are provided for removal of pivot bearing cups.

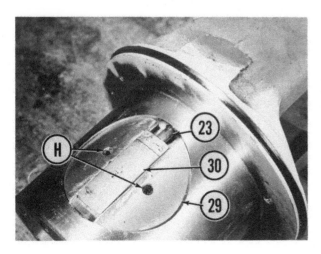

Fig. IH116 — View of pivot bearing top cap (29) showing location of knock-out holes (H). Note also adjusting wedge (30) and wedge adjusting screw (23).

lower pivot bearing can be removed from spindle by driving on a punch inserted through the knock-out holes provided in spindle. Upper pivot bearing can be removed from upper bearing cap in the same manner. See (H—Fig. IH116).

16. Clean and inspect all parts. Check bearings for roughness, damage or undue wear. Renew parts if any doubt exists as to their condition.

17. To reinstall pivot bearings, first lubricate lower bearing, drive it into position in spindle and place lower bearing cup over lower bearing. Place spindle over stub axle and support it with a jack and wood block as shown in Fig. IH117. Align bearing cup with its recess in stub axle, raise jack and press lower bearing cup into its recess. Note: If necessary, bearing cup can be seated by driving on top side of stub axle while spindle is supported. Drive upper bearing cup into its recess on top side of stub axle. Install upper pivot bearing on top cap and install assembly in spindle with shallowest end of adjusting wedge slot toward outside. Be sure adjusting wedge is installed with rounded surface upward.

Wheel bearings can now be installed and pivot bearings adjusted as outlined in paragraph 14.

FRONT DRIVE AXLE

Models So Equipped

18. **R & R AXLE ASSEMBLY.** To remove the "All-Wheel Drive" front axle and wheels as an assembly, first disconnect tie rods from steering arms. Remove "U" bolts from drive shaft front universal joint and separate universal joint. Disconnect stay rod from bracket under clutch housing and position a rolling floor jack under stay rod to support assembly. Support front of tractor, drive out pivot pin, then raise front of tractor and roll the axle and wheels assembly forward away from tractor.

19. **OVERHAUL.** Overhaul of the axle assembly can be considered as two operations and both operations can be accomplished without removing the axle assembly from the tractor. One operation concerns the stub axle along with the hub and the parts which make up the outer end of the axle. The other operation concerns the differential and carrier which is carried in the axle center housing. With the exception of removing the driving axles, which is involved in work of either operation, overhaul of either portion of the "All-Wheel Drive" axle can be accomplished without disturbing the other section.

20. **AXLE OUTER END.** Outer end of axle can be disassembled as follows: Remove wheel cover (51—Fig. IH107) and the wheel and tire assembly. Clip lock wire and remove axle retaining cap screws and tapered bushings (47) and pull axle from housing. Use caution when removing axle not to damage oil seal (21) located in axle center housing. Also see Fig. IH109. Remove the two plugs and pins (43—Fig. IH107) and remove power yoke (41) and compensating

ring (37) assembly from hub. Note: Pins (43) are tapped so cap screws can be used to aid in removal as shown in Fig. IH118. Remove roll pins and yoke pins (42—Fig. IH107) and separate power yoke and compensating ring. Straighten lock plate tabs and remove bearing adjustment plate (34), shims (24), outer wheel bearing (33) and hub (44). Note: Identify and keep removed shims in their original relationship. Loosen jam nut and wedge adjusting screw (23), then with a small punch positioned in slot of upper pivot bearing cap (29), bump adjusting wedge (30) inward until clamp ring (31) is free. See Fig. IH111. Use punch in the knock-out holes provided in spindle and bump inner wheel bearing about ½-inch toward outer end of spindle, then attach a puller, if necessary, and complete removal of bearing. Note: Be sure to keep bearing straight during removal as damage (scoring) to spindle could result. Use punch in knock-out holes (H—Fig. IH115) provided in stub axle and bump pivot bearing cups from recesses in stub axle, then tilt inner end of spindle upward and pull assembly from stub axle.

NOTE: Do not use force to remove spindle from stub axle. If spindle cannot be removed freely it is probable that one or the other pivot bearing cups are not completely free of stub axle.

If necessary, stub axle (26—Fig. IH-107) can now be removed from center housing (17).

21. Clean and inspect all parts. Pay particular attention to wheel bearings and pivot bearings in regard to roughness, damage or wear and renew any which are in any way doubtful. Wheel bearing cups can be bumped from hub, if necessary. Compensating ring bushings are available for service and renewal procedure is obvious. Be sure also that wear on driving pins (42 and 43) is not excessive. Check dowel pins in hub and roll pins in yoke pins to see that they are straight and not unduly worn.

Reassemble by reversing the disassembly procedure and adjust pivot bearings and wheel bearings as outlined in **paragraph 14.**

22. DIFFERENTIAL UNIT. To remove and overhaul the differential unit, first drain differential housing and disconnect left tie rod from left steering arm. Remove "U" bolts from front universal joint and separate universal joint. Remove both front wheel covers, then remove both front axles.

Fig. IH118 — Use cap screws as shown to pull hub to compensating ring pins.

Unbolt carrier (1—Fig. IH120) and pull carrier and differential assembly from axle center housing. See Fig. IH121.

With unit removed, remove bearing caps (2—Fig. IH120), then remove differential, bearing cups (24) and bearing adjusters (4) from carrier. Differential carrier bearings (25) can be removed from differential at this time, if necessary. Remove cap screws retaining oil seal retainer (17) and bear-

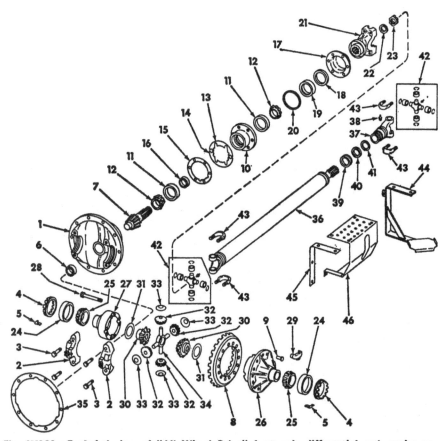

Fig. IH120—Exploded view of "All Wheel Drive" front axle differential and carrier assembly showing component parts and their relative positions.

1. Carrier	18. Felt washer	33. Thrust washer
2. Bearing cap	19. Oil seal	34. Spider
4. Bearing adjuster	20. Cork washer	35. Gasket
5. Adjuster lock	21. Yoke	36. Stub (drive) shaft
6. Pilot bearing	22. Washer	37. Sleeve yoke
7. Pinion	23. Nut	38. Grease fitting
8. Bevel gear	24. Bearing cup	39. Dust cap
9. Rivet	25. Bearing	40. Cork washer
10. Bearing cage	26. Case half	41. Steel washer
11. Bearing cup	(flanged)	42. "U" joint
12. Bearing	27. Case half (plain)	package
13. Shim (upper)	29. Lubricator	43. "U" bolt
14. Shim (lower)	30. Side gear	44. Support (Farmall)
15. Shim	31. Thrust washer	45. Shield support
16. Spacer	32. Pinion gear	(Farmall)
17. Seal retainer		46. Shield (Farmall)

ing cage (10) to carrier and remove pinion and bearing assembly. Identify and save bearing cage shims (13, 14 and 15). Remove nut (23), washer (22), yoke (21) and seal retainer (17) from pinion (7), then press pinion from outer bearing (12) and bearing cage (10). Spacer (16), inner bearing (12) and pilot bearing (6) can now be removed from pinion although pilot bearing (6) should be unstaked from pinion prior to removal. Bearing cups (11) can be driven from bearing cage (10) after pinion shaft is out. Match mark the differential case halves (26 and 27), remove case retaining bolts, then separate case halves and remove spider (34), pinions (32), thrust washers (33), side gears (30) and side gear thrust washers (31). If not already done, bearings (25) can now be removed from differential case halves. If necessary, lubricator (29) can also be removed. If bevel drive gear (8) is to be renewed, remove rivets (9) by drilling and punching.

Clean all parts and inspect for undue wear or scoring, chipped teeth or other damage and renew parts as necessary.

Reassembly is the reverse of disassembly, however, consider the following information during assembly. Shims (13 and 14) are available in 0.003 thickness. Shims (15) are available in thicknesses of 0.005, 0.010 and 0.030. Spacer (16) is available in widths of 0.506 through 0.526 in increments of 0.001. Pinion and bevel ring gear are available only as a matched set. Note also that axle oil seals (21—Fig. IH107) can be renewed when carrier and differential assembly is out.

23. To reassemble differential unit, proceed as follows: Place inner bearing cone (12) over pinion shaft (7) with largest diameter toward gear and

press bearing on shaft until it bottoms. Invert pinion shaft and press pilot bearing (6) on end of pinion shaft and stake in at least four positions. Install bearing cups (11) in bearing cage (10) with smallest inside diameters toward center. Insert pinion shaft and inner bearing in bearing cage and install spacer (16), outer bearing cone (12), seal assembly retainer (17), yoke (21), washer (22) and nut (23). Attach a holding fixture to yoke and while turning bearing cage (10), tighten nut (23) to a torque of 255 ft.-lbs. With nut tightened, check the rotation of the pinion shaft which should require a rolling torque of 8-15 in.-lbs. If shaft preload is not as stated, change the spacer (16) as required. Spacers are available in thicknesses of 0.506 through 0.526 in increments of 0.001. Use original shims (13, 14 and 15) as a starting point and install pinion shaft assembly in carrier.

Reassemble the differential assembly by reversing the disassembly procedure. Place differential in carrier, install bearing caps and bearing adjusters and tighten bearing adjusters until differential bearings have zero end play. Differential assembly can be moved left or right as required by loosening one bearing adjuster and tightening the opposite. Move the differential toward the pinion shaft until backlash is approximately 0.006-0.012. Paint ten or twelve teeth of bevel ring gear with red lead or prussian blue and turn pinion in direction of normal rotation. Tooth contact pattern should be located approximately midway of both length and width (depth) of bevel ring gear teeth. If pattern is too far toward toe of bevel gear teeth, turn adjusters as required to move gear away from pinion. If pattern is too far toward heel, turn adjusters (4) as required to move gear toward pinion. If pattern is too far toward root of bevel ring gear teeth, vary shims (13 and 14) to move pinion away from bevel drive gear. If pattern is too near top of bevel ring gear teeth, vary shims (13 and 14) to move pinion toward bevel drive gear. Continue this operation until tooth pattern is centered on the bevel ring gear teeth and backlash is 0.006-0.012 between pinion and bevel gear.

When a satisfactory pattern is obtained, slightly preload differential carrier bearings by tightening each bearing adjuster one notch. Lock both bearing adjusters.

Complete reassembly by reversing the disassembly procedure.

DRIVE HOUSING ASSEMBLY (TRANSFER CASE)

Models So Equipped

The drive housing which transmits power to the front axle is mounted on the left front side of the tractor rear frame and is driven by an idler gear on the reverse idler shaft of the range transmission.

24. REMOVE AND REINSTALL. To remove drive housing, remove shield, then remove "U" bolts of universal joint located closest to drive housing and separate the universal joint. Use tape or some other suitable means, to retain universal bearing cups on spider. Disconnect shifter rod from shifter shaft lever, drain rear frame, then unbolt and remove drive housing from tractor rear frame.

Before installing the unit on tractor, first determine the number and thickness of shim gaskets to be used between drive housing and rear frame as follows: Place a punch mark on "U" joint yoke 2 inches from center line of output shaft as shown in Fig. IH122. Using a dial indicator as shown, record the backlash between the small spool gear and output gear. Next, engage the transmission park lock and measure the backlash between park lock, reverse driven gear and reverse drive gear as shown in Fig. IH123. Record the backlash. Then, with park lock engaged, install drive housing assembly using one thin shim gasket. With dial indicator tip on the previously installed punch mark on yoke as shown in Fig. IH124, check and record the backlash. This backlash reading should be 0.003-0.007 greater than the sum of the two previous backlash readings. If the backlash increase is less than 0.003, install one thick gasket or a combination of shim gas-

Fig. IH121—View showing differential carrier and differential unit removed from center housing.

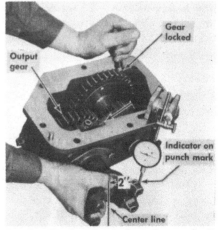

Fig. IH122—Install punch mark on yoke 2 inches from center line of output shaft. Check backlash as shown.

Fig. IH123 — With park lock engaged, check backlash between park lock, reverse driven gear and reverse drive gear.

kets to increase the backlash. Gaskets are available in two thicknesses, (thin) 0.011-0.019 and (thick) 0.0016-0.024.

Installation of drive housing unit to tractor rear main frame will be facilitated if two ½-inch guide studs are used.

25. OVERHAUL. With unit removed as outlined in paragraph 24, wedge gears and remove the esna nut which retains the universal joint yoke (28—Fig. IH125) to output shaft and remove yoke from shaft. Loosen jam nut and lock screw (24) in shifter fork and remove shifter shaft (25), spacer (27) and fork (23). Position housing with open side up, remove the roll (spring) pin (8) which retains spool gear shaft in housing, then remove spool gear shaft (5), spool gear (9) and thrust washers (12). Needle bearings in spool gear can be renewed at this time. Remove output shaft oil seal cage (20) and gasket. Lift rear snap ring (16) from its groove in output shaft and slide rearward. Slide output shaft gear (17) rearward and catch

detent balls (14) and spring (15) as they emerge from bore of output shaft. Complete removal of shaft and lift gear from housing. Ball bearing (18) can be removed from shaft after removing snap ring (19) and needle bearing (3) can be removed from bore in housing. Any further disassembly is obvious.

Clean and inspect all parts for scoring, undue wear, chipped teeth or other damage and renew as necessary. Reassemble by reversing the disassembly procedure and refer to paragraph 24 for installation information.

DRIVE SHAFT
Models So Equipped

The drive shaft between drive housing and the front axle differential is conventional and can be removed and serviced as outlined in paragraph 26.

26. R&R AND OVERHAUL. To remove drive shaft from tractor, first remove shield, then remove "U" bolts from front and rear universal joint yokes and lift shaft from tractor. The

four exposed bearing cups can now be removed from the universal joint spiders. Unscrew dust cap from sleeve yoke, pull sleeve yoke from drive shaft and remove dust cap, steel washer and cork washer. Remove bearing cup retaining snap rings from yokes, then remove bearing cups by driving spider first one way then the other.

Individual parts available for service are sleeve yoke, drive shaft, dust cap, steel washer and cork washer. Universal joint bearing cups and spider are available only as a package.

Reassembly is the reverse of disassembly, however, be certain that bearing cups are packed with grease (either by hand during assembly or through grease fitting after assembly). Bearing cups can usually be pressed into yokes simultaneously by using a bench vise. Small spacers can be used to complete installation of cups and allow installation of snap rings. Be sure ends of spider do not catch ends of bearing needles when pressing in the bearing cups.

Fig. IH124 — Install unit using one thin shim gasket and check total backlash. Refer to text.

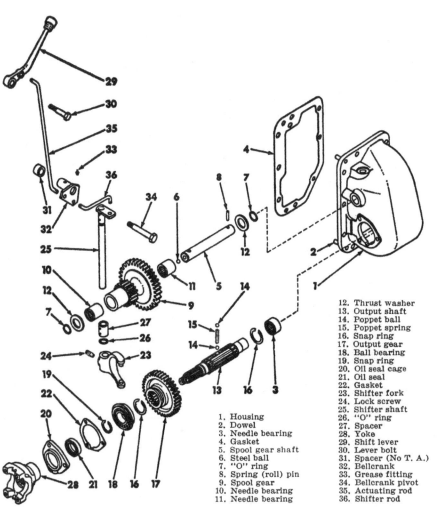

1. Housing	12. Thrust washer
2. Dowel	13. Output shaft
3. Needle bearing	14. Poppet ball
4. Gasket	15. Poppet spring
5. Spool gear shaft	16. Snap ring
6. Steel ball	17. Output gear
7. "O" ring	18. Ball bearing
8. Spring (roll) pin	19. Snap ring
9. Spool gear	20. Oil seal cage
10. Needle bearing	21. Oil seal
11. Needle bearing	22. Gasket
	23. Shifter fork
	24. Lock screw
	25. Shifter shaft
	26. "O" ring
	27. Spacer
	28. Yoke
	29. Shift lever
	30. Lever bolt
	31. Spacer (No T. A.)
	32. Bellcrank
	33. Grease fitting
	34. Bellcrank pivot
	35. Actuating rod
	36. Shifter rod

Fig. IH125—Exploded view of drive housing (transfer case) used when tractors are equipped with "All Wheel Drive".

POWER STEERING SYSTEM

NOTE: The maintenance of absolute cleanliness of all parts is of utmost importance in the operation and servicing of the hydraulic power steering system. Of equal importance is the avoidance of nicks or burrs on any of the working parts.

OPERATION

All Models

28. Power steering is standard equipment on all tractors and except for the steering cylinders and front support which differ between Farmall and International tractors, components for all tractors are similar. Refer to Figs. IH126 and IH127 for views showing the general lay-out of the component parts.

The pressurized oil used for the power steering, power brakes and the direct drive or torque amplifier clutches is furnished by a 9 gallon per minute pump located inside of the clutch housing and mounted on the inner side of the multiple control valve. Of the 9 gpm supplied by the pump, a priority of 3 gpm is taken by a flow divider located in the multiple control valve and is utilized by the power steering, brakes and torque amplifier clutches while the remaining 6 gallons is sent through the oil cooler and returned to lubricate the differential assembly and cool oil in the reservoir. The priority 3 gpm which was diverted to the power steering has the return oil flow controlled by a pressure regulator (in multiple control valve) and a second priority flow of 1 gpm is taken to operate the power brakes. The remaining 2 gpm of the original 3 gpm is diverted to the torque amplifier clutches, and as either the direct drive clutch or torque amplifier clutch is always engaged during tractor operation, the 2 gpm is available for other work so it is again controlled by a lubrication regulator (in multiple control valve) and utilized to lubricate the drive (torque amplifier) clutches and the transmission before it is returned to the reservoir.

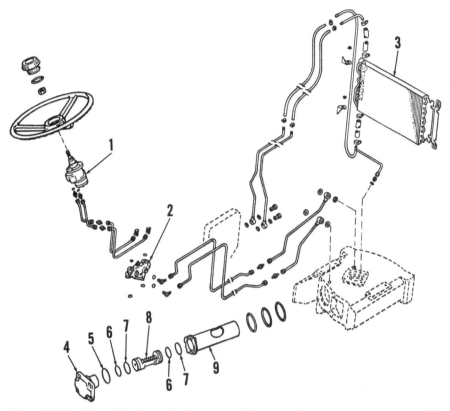

Fig. IH126—Schematic view showing general lay-out of component parts comprising power steering system on Farmall tractors.

1. Hand pump
2. Pilot valve
3. Oil cooler
4. End cover
5. "O" ring
6. "O" ring
7. Piston ring
8. Piston
9. Cylinder sleeve

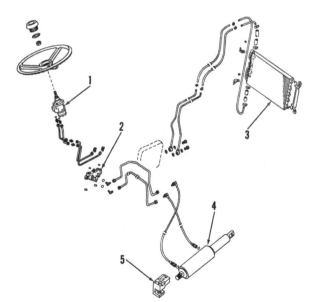

Fig. IH127 — Schematic view showing general lay-out of component parts comprising power steering system on International tractors.

1. Hand pump
2. Pilot valve
3. Oil cooler
4. Steering cylinder
5. Cylinder mounting clevis

LUBRICATION AND BLEEDING

All Models

29. The tractor rear frame serves as a common reservoir for all hydraulic and lubrication operations. The filter, shown in Fig. IH128 or IH129, should be renewed at 10 hours, 100 hours, 250 hours, and then every 250 hours thereafter. The tractor rear frame should be drained and new fluid added every 1000 hours, or once a year, whichever occurs first.

Only IH "Hy-Tran" fluid should be used and level should be maintained at the "FULL" mark shown on level gage (dip stick). The dip stick is located on the top right front of rear frame on all series.

Whenever power steering lines have been disconnected, or fluid drained,

Fig. IH128—Filter for all hydraulic operations is located on right side of tractor rear frame. Refer also to Fig. IH129.

start engine and cycle power steering system from stop to stop several times to bleed air from system, then if necessary, check and add fluid to reservoir.

TROUBLE SHOOTING

All Models

30. The following table lists some of the troubles which may occur in the operation of the power steering system. When the following information is used in conjunction with the information in the Power Steering Operational Tests section (paragraphs 32 through 40), no trouble should be encountered in locating system malfunctions.

1. No power steering or steers slowly.
   a. Binding mechanical linkage.
   b. Excessive load on front wheels and/or air pressure low in front tires.
   c. Steering cylinder piston seal faulty or cylinder damaged.
   d. Faulty power steering supply pump.
   e. Faulty commutator in hand pump.
   f. Flow divider valve spool sticking or leaking excessively.
   g. Control (pilot) valve spool sticking or leaking excessively.
   h. Circulating check ball not seating.
   i. Flow control valve orifice plugged.
2. Will not steer manually.
   a. Binding mechanical linkage.
   b. Excessive load on front wheels and/or air pressure low in front tires.
   c. Pumping element in hand pump faulty.
   d. Faulty seal on steering cylinder or cylinder damaged.
   e. Pressure check valve leaking.
   f. Control (pilot) valve spool binding or centering spring broken.
   g. Check valve in clutch housing inlet tube stuck in closed position.

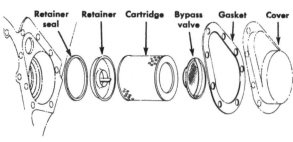

Fig. IH129 — Exploded view of hydraulic filter assembly.

3. Hard steering through complete cycle.
   a. Low pressure from supply pump.
   b. Internal or external leakage.
   c. Line between hand pump and control (pilot) valve obstructed.
   d. Faulty steering cylinder.
   e. Binding mechanical linkage.
   f. Excessive load on front wheels and/or air pressure low in front tires.
   g. Cold hydraulic fluid.
4. Momentary hard or lumpy steering.
   a. Air in power steering circuit.
   b. Control (pilot) valve sticking.
5. Shimmy.
   a. Control (pilot) valve centering spring weak or broken.
   b. Control (pilot) valve centering spring washers bent, worn or broken.

OPERATING PRESSURE AND RELIEF VALVE

All Models

31. System operating pressure and relief valve operation can be checked as follows: Remove the small orifice chamber plug (13—Fig. IH130) which is the bottom plug located on rear side of the multiple control valve and install a gage capable of registering at least 3000 psi, then start engine and operate until hydraulic fluid is warm. Run engine at high idle rpm, turn front wheels in either direction until they reach stop, then continue to apply steering effort to steering wheel so system is pressurized and note reading on the gage. Gage should read approximately 1800-1900 psi. If pressure is not as specified, renew the safety relief valve (14—Fig. IH131) located on bottom side of the multiple control valve. Relief valve is available as a unit only.

OPERATIONAL TESTS

All Models

32. The following tests are valid only when the power steering system is completely void of any air. If necessary, bleed system as outlined in paragraph 29 before performing any operational tests.

33. MANUAL PUMP. With transmission pump inoperative (engine not running), attempt to steer manually in both directions. NOTE: Manual steering with transmission pump not running will require high steering effort. If manual steering can be accomplished with transmission pump inoperative, it can be assumed that the manual pump will operate satisfactorily with the transmission pump operating.

Refer also to paragraphs 35 and 36 for information regarding steering wheel (manual pump) slip.

34. CONTROL (PILOT) VALVE. Attempt to steer manually (engine not running). Manual steering will require high steering effort but if steering can be accomplished, control (pilot) valve is working.

No steering can be accomplished if control valve is stuck on center. A control valve stuck off center will allow steering in one direction only.

35. STEERING WHEEL SLIP (CIRCUIT TEST). Steering wheel slip is the term used to describe the inability of the steering wheel to hold a given position without further steering movement. Wheel slip is generally due to leakage, either internal or external, or a faulty hand pump, steering cylinder or control (pilot) valve. Some steering wheel slip, with hydraulic fluid at operating temperature, is normal and permissible. A maximum of four revolutions per minute is acceptable. By using the steer-

Fig. IH130—Orifice chamber plug (13) is the bottom plug on rear side of multiple control valve. Also see Fig. IH131.

ing wheel slip test and a process of elimination, a faulty unit in the power steering system can be located.

However, before making a steering wheel slip test to locate faulty components, it is imperative that the complete power steering system be completely free of air before any testing is attempted.

To check for steering wheel slip (circuit test), proceed as follows: Check reservoir (rear frame) and fill to correct level, if necessary. Bleed power steering system, if necessary. Bring power steering fluid to operating temperature, cycle steering system until all components are approximately the same temperature and be sure this temperature is maintained throughout the tests. Remove steering wheel cap (monogram), then turn front wheels until they are against stop. Attach a torque wrench to steering wheel nut. NOTE: Either an inch-pound, or a foot-pound wrench may be used; however, an inch-pound wrench is recommended as it is easier to read. Advance hand throttle until engine reaches rated rpm, then apply 72 inch-pounds (6 foot-pounds) to torque wrench in the same direction as the front wheels are positioned against the stop. Keep this pressure (torque) applied for a period of one minute and count the revolutions of the steering wheel. Use same procedure and check the steering wheel slip in the opposite direction. A maximum of four revolutions per minute in either direction is acceptable and system can be considered as operating satisfactorily. If, however, the steering wheel revolutions per minute exceed four, record the total rpm for use in checking the steering cylinder or hand pump.

NOTE: While four revolutions per minute of steering wheel slip is acceptable, it is generally considerably less in normal operation.

36. MANUAL (HAND) PUMP TEST. If steering wheel slip is more than four rpm, disconnect one line between the hand pump and the control (pilot) valve. Plug the openings securely. With engine running at rated rpm apply 72 inch-pounds (6 foot-pounds) with torque wrench on steering wheel nut in direction to pressurize the plugged line. Check the steering wheel revolutions for one minute. A maximum of two rpm is acceptable and hand pump is good. If the slip is more than two rpm, repair or renew the hand pump.

37. STEERING CYLINDER TEST. If steering wheel slip, as checked in paragraph 35, exceeds the maximum of four revolutions per minute, proceed as follows: Be sure operating temperature is being maintained, then disconnect and plug the steering cylinder lines. Repeat the steering wheel slip test, in both directions, as described in paragraph 35. If steering wheel slip is ½ rpm or more, less than that recorded in paragraph 35, overhaul or renew the steering cylinder.

38. FLOW DIVIDER. When checking the flow divider operation on all models, also check the orifice (located directly below) to see that it is open and clean as this orifice is the unit that actually meters the 3 gpm used to operate the power steering.

To check operation of flow divider and orifice, proceed as follows: Disconnect one of the steering cylinder lines, and if flow rating equipment is available, attach it to the supply side of the disconnected line. If no flow rating equipment is available, use a suitable clean container to catch fluid. Start engine and run at rated rpm and keep control (pilot) valve open by applying steering effort to steering wheel in the direction of the disconnected line. Fluid flow from line should be approximately 3 gallons per minute.

If fluid flow is not as specified, service flow divider as outlined in paragraph 48. Bleed power steering system after reconnecting steering cylinder line, if necessary.

39. HYDRAULIC PUMP. To check the hydraulic pump operating pressure, refer to paragraph 31. To check the hydraulic pump free flow, proceed as outlined in paragraph 40.

40. To check pump free flow on all models, first remove the orifice plug (13—Fig. IH130 and IH131) and remove orifice (12).

NOTE: Removal of orifice will be simplified if a small wire is attached to the screw driver and extended about two inches beyond end of bit. Insert wire through orifice hole and wire will support orifice during removal from multiple control valve.

Disconnect the lower oil cooler line from front side of multiple control valve and place a container under opening to catch fluid. If available, connect a flow rater to the flow divider port and secure discharge end in the filler hole of main frame. If no flow rating equipment is available, use a short piece of hose attached to flow divider port and place discharge end in a suitable container. Run engine at rated speed and check output (free flow) of pump which should be approximately 9 gallons per minute, with a slight leakage coming from the open oil cooler port.

If pump free flow is not as specified, remove and service pump as outlined in paragraphs 41, 42 and 43.

PUMP

All Models

41. REMOVE AND REINSTALL. To remove the power steering pump, first remove bottom plug and drain clutch housing. Disconnect the "Torque-Amplifer" control rod (A—Fig. IH132) from operating bellcrank, then unbolt bracket (B). Disconnect clutch operating rod (C) at both ends and remove clutch rod and bracket (B). Disconnect brake supply line (S),

Fig. IH131—Multiple control valve with front and rear plates and pump removed and showing control valves removed from their bores.

1. Drive selector valve	8. Pin
2. Spring	9. Pressure switch
3. Brake check valve	10. Flow divider valve
4. Spring	11. Spring
5. Pressure regulator valve	12. Flow divider orifice
6. Clutch dump valve	13. Orifice plug
7. Lubrication regulator valve	14. Safety (relief) valve
	15. Oil cooler by-pass valve

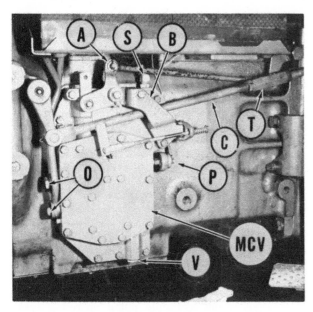

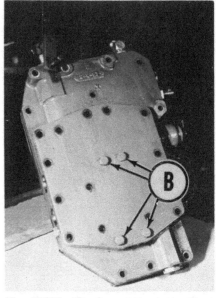

Fig. IH132 — Multiple control (MCV) valve is mounted on left front side of clutch housing.

A. Torque Amplifier control rod
B. Bracket
C. Clutch rod
O. Oil cooler lines
P. Pressure switch
S. Brake line
T. Turnbuckle
V. Safety (relief) valve

Fig. IH133—The four cap screws shown at (B) retain hydraulic pump to inner face of multiple control valve.

both oil cooler lines (O) and wire from oil pressure switch (P). Remove all the multiple control valve mounting cap screws, except the pump retaining cap screws shown in Fig. IH-133, and remove the multiple control valve and pump assembly. See Fig. IH134.

NOTE: During removal of pump assembly be very careful not to lose or damage the small check valve and spring which is located in the clutch housing as shown in Fig. IH135. This check valve allows fluid to be drawn into power steering circuit when steering with engine inoperative.

Pump can now be removed from the multiple control valve by removing the four cap screws shown in Fig. IH133.

Before reinstalling pump, measure the gasket located between multiple control valve and clutch housing with a micrometer, to determine which thickness of gasket is used. These gaskets are available in light (0.011-0.019) and heavy (0.016-0.024) thicknesses and in addition to sealing, also control backlash between pump drive gear and the driving gear (pto driven gear). When reinstalling pump and valve assembly, use new gasket of same thickness as original and reinstall assembly by reversing the removal procedure.

If necessary to adjust engine clutch, refer to paragraph 220 or 221.

42. OVERHAUL (CESSNA). With pump removed as outlined in paragraph 41, remove the esna nut which retains drive gear and remove gear and the square drive key. Remove cover (3—Fig. IH136) from body. All parts are now available for inspection and/or renewal. Items 4, 5, 6, 7 and 10 are not available separately but must be ordered in a package

which includes all the "O" rings and seals.

When reassembling, install seal for drive gear shaft with lip toward inside of pump. Use new "O" rings in mounting surface. Install seal and protector in cover in same position as original. Bronze surfaces of diaphragms (7 and 10) face pump gears. Tighten body bolts to a torque of 27-30 ft.-lbs. and fill pump with oil prior to mounting on multiple control valve.

Refer to the following table for dimensional information.

Cessna

Gear width (new)	0.444
Gear width (min. allowable)	0.441
Gear width variation (max. allowable)	0.001
Bearing I. D. (new)	0.6886
Bearing I. D. (max. allowable)	0.691
Gear shaft diameter (new)	0.6875
Gear shaft diameter (min. allowable)	0.685
Body bore I. D. (new)	1.716
Body bore I. D. (max. allowable)	1.719
Gear to body clearance (max.)	0.005

43. OVERHAUL (THOMPSON). With pump removed as outlined in paragraph 41, remove the esna nut which retains drive gear and remove gear and the square drive key. Remove cover (2—Fig. IH137) from body. All parts are now available for inspection and/or renewal. Note position of pressure plate (6) as it is removed. Items 3, 4, 5, 6 and 9 are not available separately but must be ordered in a package which includes all the "O" rings and seals.

When reassembling, install oil seal for drive gear shaft with lip toward inside of pump. Use new "O" rings in mounting surface. Install sealing

Fig. IH134 — With multiple control valve removed, the power steering pump is located as shown.

Fig. IH135 — With multiple control valve off, use caution not to lose or damage the small check valve and spring shown at (CV). Pump is driven by the pto driven gear (G).

web, web backings, pressure plate and wear plate in same position as original. Tighten body bolts to a torque of 27-30 ft.-lbs. and fill pump with oil prior to installation on multiple control valve.

Refer to the following table for dimensional information.

Thompson
Gear width (new)............0.540
Gear width (min. allowable)...0.5395
Gear width variation
(max. allowable)...........0.0005
Bearing I. D. (new)..........0.8150
Bearing I. D. (max. allowable).0.8155
Gear shaft diameter (new).....0.8125
Gear shaft diameter
(min. allowable)............0.8120
Body bore I. D. (new).......1.7710
Body bore I. D.
(max. allowable)...........1.7715
Gear to body clearance (max.).0.0029

HAND PUMP

All Models

44. REMOVE AND REINSTALL. To remove the power steering hand pump, first remove cap (monogram) from steering wheel, remove nut, attach puller and remove steering wheel. NOTE: DO NOT drive on steering wheel or shaft to remove steering wheel or damage to hand pump could occur. Remove the small plate just below steering wheel on rear side of steering wheel support and disconnect lines from hand pump. Remove button plug from hand throttle lever, then remove nut, washer and hand throttle lever from throttle control shaft. Unbolt support cover and lift cover and hand pump from support. Separate hand pump from support cover.

45. OVERHAUL MANUAL (HAND) PUMP. Remove the manual pump as outlined in paragraph 44. Clear fluid from unit by rotating steering wheel (input) shaft back and forth several times. Place unit in a soft jawed vise with end plate on top side, then remove end plate retaining cap screws and lift off end plate (2—Fig. IH138).

NOTE: Lapped surfaces of end plate (2), pumping element (5), spacer (6), commutator (9) and pump body (13) must be protected from scratching, burring or any other damage as sealing of these parts depends only on their finish and flatness.

Remove seal retainer (3), seal (4), pumping element (5) and spacer (6) from body (13). Remove commutator (9) and drive link (8), with link pins (7) and commutator pin (9A), from body. Smooth any burrs or nicks which may be present on input shaft (10), wrap spline with masking tape, then remove input shaft from body. Remove bearing race (12) and thrust bearing (11) from input shaft. Remove snap ring (19), washer (18), spacer (17), back-up washer (16) and seal (15). Do not remove needle bearing (14) unless renewal is required. If it should be necessary to renew bearing, press same out pumping element end of body.

Clean all parts in a suitable solvent and if necessary, remove paint from outer edges of body, spacer and end plate by passing these parts lightly over crocus cloth placed on a perfectly

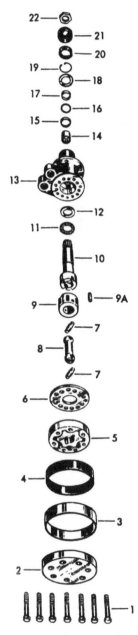

Fig. IH138—Exploded view of the power steering hand pump.

1. Cap screw	11. Thrust bearing
2. End plate	12. Bearing race
3. Seal retainer	13. Body
4. Seal	14. Needle bearing
5. Pumping element	15. Seal
6. Spacer plate	16. Back-up washer
7. Link pin	17. Spacer
8. Drive link	18. Washer
9. Commutator	19. Retainer
9A. Commutator pin	20. Felt seal
10. Coupling (Input)	21. Water seal
shaft	22. Nut

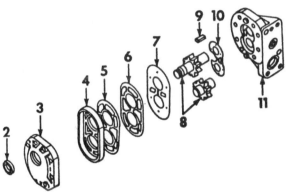

Fig. IH136 — Exploded view of Cessna power steering pump.

2. Seal
3. Cover
4. Diaphragm seal
5. Protector gasket
6. Back-up gasket
7. Diaphragm
8. Gears and shafts
9. Key
10. Body diaphragm
11. Pump body

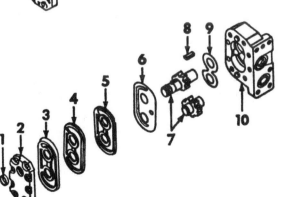

Fig. IH137 — Exploded view of Thompson power steering pump. Note position of pressure plate (6).

1. Seal
2. Cover
3. Sealing web
4. Backing web (paper)
5. Backing web
6. Pressure plate
7. Pump gears and shafts
8. Drive key
9. Wear plate
10. Pump body

flat surface. Do not attempt to dress out any scratches or other defects since these sealing surfaces are lapped to within 0.0002 of being flat.

Inspect commutator and housing for scoring and undue wear. Bear in mind that burnish marks may show, or discolorations from oil residue may be present on commutator after unit has been in service for some time. These can be ignored providing they do not interfere with free rotation of commutator in body.

Check fit of commutator pin in the commutator. Pin should be a snug fit and if bent, or worn until diameter at contacting points is less than 0.2485, renew pin.

Measure inside diameter of input shaft bore in body and outside diameter of input shaft bearing surface. If body bore is 0.006 or more larger than shaft diameter, renew shaft and/or body and commutator. Note: Body and commutator are not available separately.

Check thrust bearing and race for excessive grooving, flat spots or any other damage and renew bearing assembly if necessary.

Place pumping element on a flat surface and in the position shown in Fig. IH139. Use a feeler gage and check clearance between ends of rotor teeth and high points of stator. If clearance exceeds 0.003, renew pumping element. Use a micrometer and measure width (thickness) of rotor and stator. If stator is 0.002 or more wider (thicker) than the rotor, renew the pumping element. Pumping element rotor and stator are available only as a matched set.

Check end plate for wear, scoring and flatness. Do not confuse the polish pattern on end plate with wear. This pattern, which results from rotor rotation, is normal. Renew end plate if worn or scored and if not within 0.0002 of being flat.

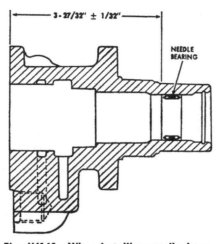

Fig. IH140—When installing needle bearing in body, install same to dimension shown.

When reassembling, use all new seals and back-up washers. All parts, except those noted below, are installed dry. Reassemble as follows: If needle bearing (14) was removed, lubricate with IH Hy-Tran fluid, install from pumping element end of body and press bearing into bore until inside end measures $3\frac{13}{16}$-$3\frac{7}{8}$ inches from pumping element end of body as shown in Fig. IH140. Lubricate thrust bearing assembly with IH Hy-Tran fluid and install assembly on input shaft with race on top side. Install input shaft and bearing assembly in body and check for free rotation. Install a link pin in one end of the drive link, then install drive link in input shaft by engaging the flats on link pin with slots in input shaft. Use a small amount of grease to hold commutator pin in commutator, then install commutator and pin in body while engaging pin in one of the long slots of the input shaft. Commutator is correctly installed when edge of commutator is slightly below sealing surface of body. Clamp body in a soft jawed vise with input shaft pointing downward. Again make sure surfaces of spacer, pumping element, body and end plate are perfectly clean, dry and undamaged. Place spacer on body and align screw holes with those of body.

Put link pin in exposed end of drive link, then install pumping element rotor while engaging flats of link pin with slots in rotor. Position pumping element stator over rotor and align screw holes of stator with those of spacer and body. Lubricate pumping element seal lightly with IH Hy-Tran fluid and install seal in seal retainer, then install seal and retainer over pumping element stator. Install end cap, align screw holes of end cap with those in pumping element, spacer and body, then install cap screws. Tighten cap screws evenly and to a torque of 18-22 ft.-lbs.

NOTE: If input shaft does not turn evenly after cap screws are tightened, loosen and retighten them again. However, bear in mind that the unit was assembled dry and some drag is normal. If stickiness or binding cannot be eliminated, disassemble unit and check for foreign material, nicks or burrs which could be causing interference.

Lubricate input shaft seal with IH Hy-Tran fluid and with input shaft splines taped to protect seal, install seal, back-up washer, spacer, washer and snap ring. The felt washer and water seal may be installed at this time but there will be less chance of loss or damage if installation is postponed until just before the steering wheel is installed.

After unit is assembled, turn unit on side with hose ports upward. Pour unit full of IH Hy-Tran fluid and work pump slowly until interior (pumping element) is thoroughly coated. Either plug ports or drain excess oil.

Reinstall unit by reversing the removal procedure and bleed steering system as outlined in paragraph 29.

PILOT VALVE

All Models

46. REMOVE AND REINSTALL. To remove the power steering control (pilot) valve, which is located on top side of clutch housing, it will be necessary to remove left battery and battery tray on diesel models. See Fig.

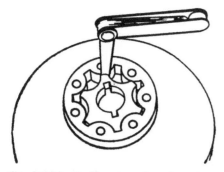

Fig. IH139—Position pumping element as shown to check tooth clearance. Refer to text.

Fig. IH141 — Installed view of power steering control (pilot) valve.

IH141. Disconnect the hand pump lines and the steering cylinder lines, then remove the three mounting cap screws and lift valve from clutch housing.

When reinstalling, renew "O" rings located between pilot valve and clutch housing. Bleed power steering system as outlined in paragraph 29.

All Models

47. OVERHAUL. With valve assembly removed, disassemble as follows: Refer to Fig. IH142 and remove end caps (1) with "O" rings (2). Pull spool and centering assembly from valve body (9). Place a punch or small rod in hole of centering spring screw (3) and remove screw, centering spring (5) and centering spring washers (4). Remove plug (11), "O" ring (12) and circulating check ball (10). Remove retainer (16), seat (15), pressure check valve (14) and spring (13).

NOTE: Some early production valves were equipped with a safety relief valve instead of plug (7). If control (pilot) valve is so

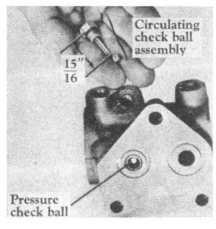

Fig. IH143—Distance between gasket surface of circulating check valve plug and inner end of roll pin should be 15/16-inch as shown.

equipped, do not disassemble the safety relief valve unless it is deemed absolutely necessary. Valve is set at factory and in normal operation is seldom actuated. Therefore, it is unlikely that valve will be damaged. However, should it be necessary to disassemble valve, count and record the number of exposed threads on the adjusting screw and be sure to reinstall the adjusting screw to the same position.

Wash all parts in a suitable solvent and inspect. Valve spool and spool bore in body should be free of scratches, scoring or excessive wear. Spool should fit its bore with a snug fit and yet move freely with no visible side play. If spool or spool bore is defective, renew complete valve assembly as spool (6) and valve body (9) are not available separately.

Inspect pressure check valve and seat. Renew parts if grooved or scored.

Reassembly is the reverse of disassembly and the following points should be observed. Coat all parts with Hy-Tran fluid, or its equivalent, prior to installation. If safety relief valve was disassembled, on valves so equipped, be sure the adjusting screw is installed with the same number of threads exposed as were exposed prior to removal. Install spool with spring end opposite circulating check ball. Measure distance between gasket surface of circulating check ball plug and end of roll pin as shown in Fig. IH143. This distance should be 15/16-inch and if necessary, obtain this measure by adjusting roll pin in or out. Tighten end cap retaining cap screws to a torque of 186 in.-lbs.

Reinstall valve by reversing removal procedure and bleed power steering system as outlined in paragraph 29.

FLOW DIVIDER

All Models

48. R&R AND OVERHAUL. The flow divider valve is located in rear side of the multiple control valve assembly and is under the second from bottom plug. See Fig. IH131. Valve (spool) can be removed after removing plug and spring.

Service of flow divider valve consists of renewing parts. Carefully inspect spool and bore for scratches, grooves or nicks. Spool should fit bore snugly yet be free enough to slide easily in bore with both spool and bore lubricated. Flow divider spring (11) has a free length of 4.08 inches and should test 33.66 pounds when compressed to a length of 2.50 inches.

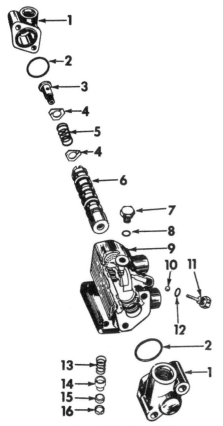

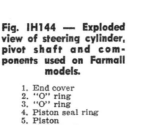

Fig. IH144 — Exploded view of steering cylinder, pivot shaft and components used on Farmall models.

1. End cover
2. "O" ring
3. "O" ring
4. Piston seal ring
5. Piston
6. Cylinder sleeve
7. "O" rings
8. Upper cover
9. Nut
10. Bearing cone
11. Bearing cup
12. "O" ring
13. Front support
14. "O" ring
15. Bearing cup
16. Bearing cone
17. Oil seal
18. Pivot shaft

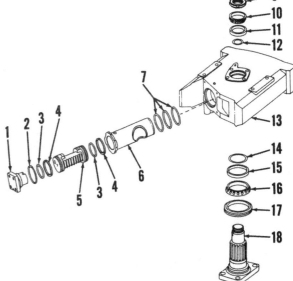

Fig. IH142—Exploded view of the power steering control (pilot) valve.

1. End cap
2. "O" ring
3. Centering spring screw
4. Centering spring washer
5. Centering spring
6. Spool
7. Plug
8. "O" ring
9. Valve body
10. Steel ball
11. Plug assembly
12. "O" ring
13. Spring
14. Check valve
15. Seat
16. Retainer

STEERING CYLINDER

Farmall Models

49. REMOVE AND REINSTALL. On Farmall models, the power steering cylinder and piston is incorporated into the front support and removal of cylinder requires removal of the front support.

To remove front support, first remove hood, then remove radiator as outlined in paragraph 200. Support tractor and on tricycle models, remove wheels and pedestal from steering pivot shaft. On adjustable wide axle models, unbolt axle support from front support, center steering arm from pivot shaft and stay rod ball from stay rod support, then roll complete assembly away from tractor. On all models, disconnect the power steering lines from front support, then attach hoist to front support, unbolt from side rails and remove from tractor. Note: If necessary, loosen engine front mounting bolts to provide additional removal clearance for front support.

Reinstall by reversing removal procedure and bleed power steering system as outlined in paragraph 29.

50. OVERHAUL. Before disassembling power steering cylinder, match mark the cylinder flange and front support. The cylinder sleeve (6—Fig. IH144) is bored off-center and the cam action resulting from rotating the cylinder will regulate the backlash between the teeth of pivot shaft (18) and teeth of piston (5). If the backlash prior to disassembly is satisfactory, the marks installed will provide the correct backlash during reassembly. If backlash is not correct, a point of reference will be established to simplify the backlash adjustment. Backlash should be adjusted to as near zero as possible without binding.

To remove the power steering cylinder and piston, first remove cap from top of pivot shaft, then straighten lock on nut, remove nut from top end of pivot shaft and pull pivot shaft from bottom of front support. Remove cylinder end cover and pull cylinder and piston from upper bolster. Piston can now be removed from cylinder. Any further disassembly required is obvious.

Reassembly is the reverse of disassembly. Renew all "O" rings and seals. When installing pivot shaft be sure to position the marked center tooth of piston rack between the two punch marked teeth of the pivot shaft as shown in Fig. IH145. Tighten pivot shaft nut to provide a slight pre-load

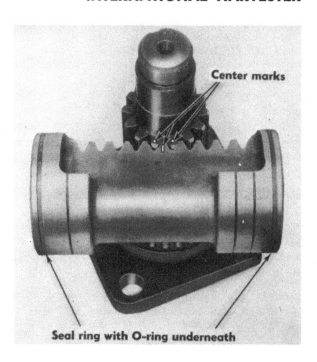

Fig. IH145—Pivot shaft and piston are correctly meshed when timing marks align as shown.

Center marks

Seal ring with O-ring underneath

on pivot shaft and secure by bending lock flange on nut into notch in pivot shaft. Lobes are provided on flange of cylinder so mesh position of pivot shaft and piston can be adjusted to zero backlash (at straight ahead position) by using a small punch and hammer.

International Models

51. R&R AND OVERHAUL. To remove the power steering cylinder on International models, first disconnect lines from cylinder and plug lines to prevent oil drainage. Disconnect steering cylinder from center steering arm and rear mounting clevis and remove from tractor.

52. With cylinder removed, move piston rod back and forth several times to clear oil from cylinder. Refer

to Fig. IH146 and proceed as follows: Place barrel of cylinder in a vise and clamp vise only enough to prevent cylinder from turning. Turn cylinder head (11) until free end of retaining ring (10) appears in slot of barrel (4). Lift end of ring to outside of barrel and continue turning head until all of ring is outside barrel, lift nib from hole and remove the ring. Remove cylinder head assembly from piston rod, then pull piston rod and piston (7) from cylinder. All seals, "O" rings and back-up washers are now available for inspection and/or renewal.

Clean all parts in a suitable solvent and inspect. Check cylinder for scoring, grooving and out-of-roundness. Light scoring can be polished out by using a fine emery cloth and oil pro-

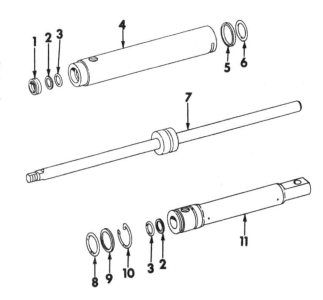

Fig. IH146 — Exploded view of the power steering cylinder used on International model tractors.

1. Seal
2. Back-up washer
3. "O" ring
4. Cylinder barrel
5. Piston ring
6. "O" ring
7. Piston rod
8. "O" ring
9. Back-up washer
10. End retainer
11. Cylinder head

Fig. IH147 — Multiple control valve (MCV) is mounted on left front side of clutch housing.

A. Torque Amplifier control rod
B. Bracket
C. Clutch rod
O. Oil cooler lines
P. Pressure switch
S. Brake line
T. Turnbuckle
V. Safety (relief) valve

viding a rotary motion is used during the polishing operation. A cylinder that is heavily scored or grooved, or that is out-of-round, should be renewed. Check piston rod and cylinder for scoring, grooving and straightness. Polish out very light scoring with fine emery cloth and oil, using a rotary motion. Renew rod and piston assembly if heavily scored or grooved, or if piston rod is bent. Inspect piston seal for frayed edges, wear and imbedded dirt or foreign particles. Renew seal if any of the above conditions are found. NOTE: Do not remove the "O" ring (8) located under the piston seal unless renewal is indicated. Inspect balance of "O" rings, back-up washers and seals and renew same if excessively worn. Be sure air bleed holes in the cylinder head assembly are open and clean.

Reassemble steering cylinder as follows: Lubricate piston seal and cylinder head "O" rings and using a ring compresser, or a suitable hose clamp, install piston and rod assembly into cylinder. Install cylinder head in cylinder so hole in cylinder head will accept nib of retaining ring and pull same into its groove by rotating cylinder head. Complete balance of reassembly by reversing the disassembly procedure.

Reinstall unit on tractor, then fill and bleed the power steering system as outlined in paragraph 29.

OIL COOLER

All Models

53. **R & R AND OVERHAUL.** An oil cooler, shown in Fig. IH126 or IH127, is incorporated into the power steering system. Of the 9 gpm supplied by the power steering pump, only 3 gpm is used to operate the power steering and tractor controls. The remaining 6 gpm is directed, via the flow divider valve, to the oil cooler where it is cooled and returned to lubricate the tractor differential and provide cooling for oil in the reservoir. Pressure regulation of the oil to oil cooler is controlled by the oil cooler by-pass valve in the multiple control valve. See Fig. IH131.

Service of the oil cooler involves only removal and reinstallation, or renewal of faulty units. Removal of the oil cooler is obvious after removal of the radiator grille and hood and an examination of the unit. However, outlet and inlet hoses must be identified as they are removed from oil cooler pipes so oil circuits will be kept in the proper sequence.

WITHOUT "TA" | **WITH "TA"**

Fig. IH148—Exploded view of the multiple control valve used on all models.

1. Pump drive gear
2. Pump
3. "O" ring
4. "O" ring
5. Gasket (heavy)
6. Gasket (light)
11. Inner plate (with TA)
12. Inner gasket
13. "O" ring
14. "O" ring
15. Body
16. Elbow
17. "O" ring
18. Spring
19. Brake check valve
20. Spring, outer
21. Pressure regulator valve
22. "O" ring
23. Plug

24. Plug
25. "O" ring
26. By-pass valve
27. Lubrication regulator spring
28. Flow control valve
29. Spring
30. "O" ring
31. Plug
32. Plug
33. "O" ring
34. Orifice
35. Gasket
36. Outer cover (plate)
37. Plug
38. "O" ring
39. Safety valve
40. "O" ring
41. "O" ring
42. Plug

43. Oil cooler by-pass spring
45. Inner gasket
47. Retaining ring
48. "O" ring
49. Valve retaining screw
50. "O" ring
51. Valve stem
52. Spring, outer
53. Spring, inner
54. Valve guide
55. "O" ring
56. Plug
57. "O" ring
58. Spring
59. Clutch dump valve
60. Body
61. Inner plate (no TA)
62. Spring, inner

MULTIPLE CONTROL VALVE

All Models

54. **R&R AND OVERHAUL.** The multiple control valve is mounted on left front of clutch housing as shown in Fig. IH147. The multiple control valve has a 9 gpm pump mounted on inner side which furnishes pressurized oil for operation of power steering, brakes and torque amplifier clutches as well as providing lubrication for the torque amplifier assembly and speed transmission assembly. The multiple control valve also contains the spools, valves and passages necessary to control these operations.

When servicing the multiple control valve, cleanliness is of the utmost importance as well as the avoidance of any nicks or burrs. When reinstalling the multiple control valve and pump assembly, be sure to use the same thickness gasket (5 or 6—Fig. IH148) as original. Use a micrometer to measure gasket. Gaskets are available in light (0.011-0.019) and heavy (0.016-0.024) thicknesses.

To remove and reinstall the multiple control valve, refer to paragraph 41.

With unit removed, all spools, plungers and springs can be removed and the procedure for doing so is obvious. However, note that before both by-pass valves (26—Fig. IH148) can be removed, the small retaining pins must be removed. Pins are retained in position by the inner and outer plates.

Refer to the following table for spring test data and to Fig. IH148 for spring location. Spring call-out numbers are in parenthesis.

Brake Check Valve (18)
    Free length—inches....0.500
    Test length lbs. @ in...0.5 @ 0.300
Pressure Regulator Valve (20 & 62)
    Outer spring free length—
      inches 3.508
    Test length lbs. @ in...64.5 @ 2.170
    Inner spring free length—
      inches 2.910
    Test length lbs @ in. ..19.5 @ 1.940
Lubrication Regulator Valve (27)
    Free length—inches....1.072
    Test length lbs. @ in...5.2 @ 0.787
Flow Divider (29)
    Free length—inches....4.08
    Test length lbs. @ in...33.7 @ 2.50
Oil Cooler By-Pass Valve (43)
    Free length—inches....1.244
    Test length lbs. @ in...26 @ 0.712
Drive Selector Valve (52 & 53)
    Outer spring free length
      —inches 1.578
    Test length lbs. @ in. ..28 @ 1.281
    Inner spring free length
      —inches 1.880
    Test length lbs. @ in...18 @ 1.254
Clutch Dump Valve (58)
    Free length—inches....1.720
    Test length lbs. @ in....5 @ 1.150

Reassembly is the reverse of disassembly. Use all new "O" rings and refer to paragraph 232 when adjusting "Torque-Amplifier" drive selector and to paragraph 225 or 226 when adjusting the "Torque-Amplifier" dump valve.

Safety relief valve (V—Fig. IH147) is heavily staked and cannot be disassembled. Valve should relieve at approximately 1800-1900 psi and faulty valves are renewed as a unit. Refer to paragraph 31 for information on checking valve.

Fig. IH149—Exploded view of tilt steering wheel assembly available for all models.

1. Cover
2. Shield
3. Back-up shim
4. Position quadrant
5. Universal joint
6. Support
7. Release lever
8. Dust seal
9. Spacer hub
10. Steering hand pump

TILT STEERING WHEEL

All Models

55. The tilt steering wheel assembly (Fig. IH149) is available for all models. To adjust steering wheel position, move release lever (7) rearward, tilt steering wheel to desired position, then allow the spring to engage release lever in notch on quadrant (4). Steering wheel can be tilted to five different positions. Disassembly and repair of the unit will be obvious after an examination of the assembly and reference to Fig. IH149.

ENGINE AND COMPONENTS (NON-DIESEL)

Farmall 706 non-diesel tractors prior to serial number 37237 and International 706 and 2706 non-diesel tractors prior to serial number 5274 are equipped with engines having a bore and stroke of $3\frac{9}{16}$ x 4 25/64 inches and a displacement of 263 cubic inches. Later series 706 and 2706 and all series 756 and 2756 non-diesel tractors are equipped with engines having a bore and stroke of 3¾ x 4 25/64 inches and a displacement of 291 cubic inches.

Series 806, 2806, 856, and 2856 non-diesel tractors are equipped with engines having a bore and stroke of 3 13/16 x 4 25/64 inches and a displacement of 301 cubic inches.

All engines are a six cylinder design. The 263 and 291 cubic inch engines are fitted with dry type cylinder sleeves and the 301 cubic inch engine is of the sleeveless type. Engines may be equipped with either gasoline or L-P gas fuel system. Dry type air filters are used in the air induction systems of all engines.

R&R ENGINE WITH CLUTCH

All Non-Diesel Models

56. Removal of engine is best accomplished by removing front support, radiator, front axle and wheels assembly and the fuel tank assembly, then unbolting and removing the engine and side rails from clutch housing as follows:

Drain cooling system, disconnect battery cables and remove hood. Disconnect radiator hoses and air cleaner inlet hose. Identify and disconnect power steering lines and hydraulic oil cooler lines. Remove the operator assist handles, torque amplifier lever and notice that TA lever and shaft have match marks affixed to insure correct assembly. Remove center hood and steering support cover. Disconnect all interfering lines and wiring from fuel tank and support. Attach hoist to tank, unbolt and lift fuel tank and support from tractor.

Disconnect remaining engine controls and electrical wiring. Support tractor under clutch housing, attach hoist to front support, unbolt front support from side rails, disconnect stay rod ball, if so equipped, and roll complete front assembly forward from tractor. NOTE: If additional clearance is needed for removal of front support from side rails, loosen the engine front mounting cap screws.

Attach hoist to engine, then unbolt and remove engine and side rails from clutch housing.

When reinstalling engine assembly, unbolt and remove clutch assembly from flywheel. Place pressure plate assembly and drive disc on shafts in clutch housing. Clutch can be bolted to flywheel after engine is installed by working through the opening at bottom of clutch housing.

CYLINDER HEAD

All Non-Diesel Models

57. **REMOVE AND REINSTALL.** To remove the cylinder head, first remove fuel tank assembly as outlined in paragraph 56. Drain cooling system and remove the coolant temperature bulb. Remove air cleaner and bracket assembly. Disconnect spark plug wires, unbolt coil bracket from cylinder head and lay coil and wiring harness out of the way. NOTE: If equipped with Magnetic Pulse ignition system, remove pulse amplifier and lay it aside. Unbolt and remove fan assembly. Disconnect upper radiator hose and bypass hose. Disconnect controls from carburetor, then unbolt and remove manifold and carburetor assembly. Remove rocker arm cover, rocker arms and shaft assembly and push

rods. Remove cylinder head cap screws and lift off cylinder head.

Use guide studs when reinstalling cylinder head, use new head gasket and make certain gasket sealing surfaces are clean. Tighten cylinder head retaining cap screws to a torque of 85-95 ft.-lbs. using the tightening sequence shown in Fig. IH150. Manifold retaining cap screws should be tightened to a torque of 45 ft.-lbs. Adjust valve tappet gap to 0.027 as outlined in paragraph 58.

VALVES AND SEATS

All Non-Diesel Models

58. Inlet and exhaust valves are not interchangeable. Inlet valves seat directly in the cylinder head; whereas, the cylinder head is fitted with renewable seat inserts for the exhaust valves. Inserts are available in standard size as well as oversizes of 0.015 and 0.030. Valve face and seat angle for both the inlet and the exhaust is 30 degrees. Valve rotators (Rotocoils) are used on exhaust valves and umbrella type stem seals are used on inlet valves.

When removing valve seat inserts, use the proper puller. Do not attempt to drive chisel under insert or counterbore will be damaged. Chill new seat insert with dry ice or liquid Freon prior to installing. When new insert is properly bottomed, it should be 0.008-0.030 below edge of counterbore. After installation, peen the cylinder head material around the complete outer circumference of the valve seat insert. The O.D. of new standard insert is 1.5655 and the insert counterbore I. D. is 1.5625.

Check the valves and seats against the specifications which follow:

Inlet

Face and seat angle ....30°

Stem diameter 0.3715-0.3725
Stem to guide
  diametral clearance 0.003-0.005
Seat width 0.048-0.074
Valve run-out (max.) 0.002
Valve tappet gap (warm) 0.027
Valve head margin 5/64
Exhaust
Face and seat angle ....30°
Stem diameter 0.371-0.372
Stem to guide
  diametral clearance ..0.0035-0.0055
Seat width 0.083-0.103
Valve run-out (max.) 0.002
Valve tappet gap (warm) 0.027
Valve head margin 5/64

To adjust valve tappet gap, crank engine to position number one piston at top dead center of compression stroke. Adjust the six valves indicated on the chart in Fig. IH151. Turn engine one revolution to position number six piston at TDC (compression) and adjust the remaining six valves indicated on chart.

VALVE GUIDES AND SPRINGS

All Non-Diesel Models

59. The inlet and exhaust valve guides are interchangeable. Inlet guides are pressed into cylinder head until top of guide is $1\frac{3}{16}$ inches above the spring recess of the cylinder head. The exhaust valve guides are pressed into cylinder head until top of guide is ¾-inch above the spring recess of head. Guides are pre-sized and if carefully installed, should require no final sizing. Inside diameter of valve guides is 0.3755-0.3765 and valve stem to guide diametral clearance is 0.003-0.005 for inlet valves and 0.0035-0.0055 for exhaust valves.

Inlet and exhaust valve springs are interchangeable. Springs should have a free length of $2\frac{7}{16}$ inches and should test 146-156 pounds when compressed to a length of $1\frac{19}{32}$ inches. Renew any spring which is rusted, discolored or does not meet the pressure test specifications.

VALVE TAPPETS
(CAM FOLLOWERS)

All Non-Diesel Models

60. Tappets are of the barrel type and operate in unbushed bores in the crankcase. Tappet diameter should be 0.9965-0.9970 and should operate with a clearance of 0.002-0.004 in the 0.9990-1.0005 crankcase bores. Tappets are available in standard size only and can be removed from side of crankcase after removing rocker arm cover, rocker arms and shaft assembly, push rods and engine side cover.

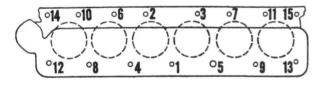

Fig. IH150 — Tightening sequence for non-diesel cylinder head cap screws.

	ADJUST VALVES (ENGINE WARM)											
With No. 1 Piston at T.D.C. (Compression)	1	2	3		5		7		9			
With No. 6 Piston at T.D.C. (Compression)				4		6		8		10	11	12

← Front   Rear →

Numbering sequence of valves which correspond to chart

Fig. IH151—Chart shows the valve tappet gap adjusting procedure for all non-diesel engines.

VALVE ROCKER ARM COVER

All Non-Diesel Models

61. Removal of the rocker arm is obvious after removal of front hood. On models 806, 2806, 856 and 2856, left front fuel tank support must be removed. When reinstalling cover, use new gasket to insure an oil tight seal.

VALVE TAPPET LEVERS (ROCKER ARMS)

All Non-Diesel Models

62. **REMOVE AND REINSTALL.** Removal of the rocker arms assembly may be accomplished in some cases without removing the fuel tank. The rocker shaft bracket hold down screws are also the left row of cylinder head cap screws and removal of these screws may allow the left side of cylinder head to raise slightly and damage to the head gasket could result. Because of possible damage to the head gasket, it is recommended that the cylinder head be removed as in paragraph 57 when removing the rocker arm assembly.

63. **OVERHAUL.** The two-piece rocker arm shaft has an outside diameter of 0.748-0.749. Rocker arm bushings have an inside diameter of 0.7505-0.7520. Bushings are not renewable. All rocker arms are interchangeable. A shaft indexing roll pin (RP—Fig. IH152) and a bracket plug are installed in the front and rear rocker shaft brackets. Numbers 2, 4 and 6 rocker shaft brackets are equipped with dowel sleeves for alignment.

VALVE ROTATORS

All Non-Diesel Models

64. Positive type valve rotators (Rotocoils) are installed on the exhaust valves of all non-diesel engines. Normal servicing of the valve rotators consists of renewing the units. It is

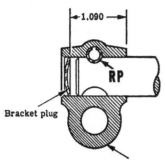

Fig. IH152—Roll pin (RP) is installed with the slot away from the rocker shaft and should engage the smaller notch in the shaft. Note installation dimension of valve lever shaft bracket plug.

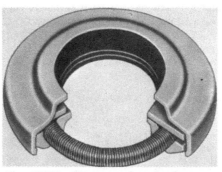

Fig. IH153—Cut-away view showing construction of a "Rotocoil" valve rotator.

important to observe the valve action after engine is started. Valve rotator action can be considered satisfactory if valve rotates a slight amount each time the valve opens. See Fig. IH153.

VALVE TIMING

All Non-Diesel Models

65. To check valve timing, remove rocker arm cover and crank engine to position number one piston at TDC of compression stroke. Adjust number one intake valve tappet gap to 0.034. Place a 0.004 feeler gage between valve lever and valve stem of number one intake. Slowly rotate crankshaft in normal direction until valve lever becomes tight on feeler gage. At this point, number one intake valve will start to open and timing pointer should be within the range of 5 to 11 degrees before top dead center.

NOTE: One tooth "out of time" equals approximately 13 degrees.

Readjust number one intake valve tappet gap as outlined in paragraph 58.

TIMING GEAR COVER

All Non-Diesel Models

66. To remove the timing gear cover, drain cooling system, remove hood and disconnect battery cables. On Farmall models, identify and disconnect power steering cylinder lines. On International models, disconnect forward end of power steering cylinder from center steering arm. Then, on all models, identify and disconnect the hydraulic oil cooler lines. Disconnect air cleaner inlet hose and radiator hoses. Support tractor under clutch housing. On Farmall models equipped with wide front axle and all International models, disconnect stay rod ball. On all models, attach hoist to front support, unbolt front support from side rails and roll complete front assembly forward from tractor.

Unbolt and remove fan, generator or alternator and drive belts. Remove

crankshaft nut, attach a suitable puller and remove crankshaft pulley. Remove cap screws retaining oil pan to timing gear cover and loosen the remaining oil pan cap screws. Unbolt timing gear cover, then pull cover forward off the dowels and remove from engine. Use care not to damage oil pan gasket.

Reassemble by reversing the disassembly procedure. Tighten crankshaft pulley retaining nut to a torque of 95 ft.-lbs.

TIMING GEARS

All Non-Diesel Models

67. **CRANKSHAFT GEAR.** Crankshaft gear is keyed and press fitted on the crankshaft. The gear can be removed using a suitable puller after first removing the timing gear cover as outlined in paragraph 66.

Before installing, heat gear in oil, then drift heated gear on crankshaft. Make certain timing marks are aligned as shown in Fig. IH154.

68. **CAMSHAFT GEAR.** Camshaft gear is keyed and press fitted on the camshaft. Backlash between camshaft gear and crankshaft gear should be 0.0032-0.0076. Camshaft gear can be removed using a suitable puller after first removing the timing gear cover as outlined in paragraph 66 and the gear retaining nut.

Before installing, heat gear in oil until gear will slide on shaft. Install lock and nut, then tighten nut to a torque of 110-120 ft.-lbs. Make certain timing marks are aligned as shown in Fig. IH154.

69. **IDLER GEAR.** To remove the idler gear, first remove the timing gear cover as outlined in paragraph 66. Idler gear shaft is attached to front of engine by a cap screw.

Idler gear shaft diameter should be 2.0610-2.0615 and clearance between shaft and renewable bushing in gear should be 0.0015-0.0045. End clearance

Fig. IH154—Gear train and timing marks on non-diesel engines.

CA. Camshaft gear   G. Governor gear
CR. Crankshaft gear I. Idler gear

of gear on shaft should be 0.009-0.013. Make certain that oil passage in shaft is open and clean.

When reinstalling, make certain that dowel on shaft engages hole in engine front plate. The shaft retaining cap screw should be torqued to 85-90 ft.-lbs.

CAMSHAFT AND BEARINGS

All Non-Diesel Models

70. **CAMSHAFT.** To remove the camshaft, first remove timing gear cover as outlined in paragraph 66. Remove rocker arms assembly and push rods. Drain and remove oil pan and oil pump. Remove engine side cover and lift out cam followers (tappets). Working through openings in camshaft gear, remove the camshaft thrust plate retaining cap screws. Carefully withdraw camshaft from engine.

Recommended camshaft end play of 0.002-0.010 is controlled by the thrust plate.

Check camshaft journal diameter against the values which follow:

No. 1 (front) 2.109-2.110
No. 2 ....2.089-2.090
No. 3 ....2.069-2.070
No. 4 ....1.4995-1.5005

When installing the camshaft, reverse the removal procedure and make certain timing marks are aligned as shown in Fig. IH154. Tighten camshaft thrust plate cap screws to a torque of 38 ft.-lbs. Refer to paragraph 82 for oil pump installation.

71. **CAMSHAFT BEARINGS.** To remove the camshaft bearings, first remove the engine as in paragraph 56 and camshaft as in paragraph 70. Unbolt and remove clutch, flywheel and the engine rear end plate. Remove expansion plug from behind camshaft rear bearing and remove the bearings.

Using a closely piloted arbor, install the bearings so that oil holes in bearings are in register with oil holes in crankcase. The chamfered end of the bearings should be installed towards the rear.

Camshaft bearings are pre-sized and if carefully installed should need no final sizing. Camshaft bearing journals should have a diametral clearance in the bearings of 0.0005-0.005.

When installing plug at rear camshaft bearing, use sealing compound on plug and bore.

ROD AND PISTON UNITS

All Non-Diesel Models

72. Connecting rod and piston assemblies can be removed from above

after removing the cylinder head as outlined in paragraph 57 and the oil pan.

Cylinder numbers are stamped on the connecting rod and cap. Numbers on rod and cap should be in register and face towards the camshaft side of engine. The arrow or front marking on top of piston should be toward front of tractor.

Two types of connecting rod bolts may be used. The PLACE bolt has a head that is either notched or concave and the shank and thread diameter are nearly the same. This type attains its tension by bending the bolt head and should be torqued to 50 ft.-lbs. The PITCH bolt has a standard bolt head with a washer face. The thread diameter is larger than the shank. This type attains its tension by stretching of the shank and should be torqued to 45 ft.-lbs.

PISTONS AND RINGS

Series 806-2806-856-2856 Non-Diesel

73. The cam ground pistons operate directly in block bores and are available in standard size as well as oversizes of 0.010, 0.020, 0.030 and 0.040. Series 806 and 2806 non-diesel tractors with engine serial numbers prior to 4997 are equipped with pistons having three compression rings and one oil control ring. Later production 806 and 2806 non-diesel tractors and all 856 and 2856 non-diesel tractors are equipped with pistons having two compression rings and one oil control ring.

Check pistons and rings against the values which follow:

Early 806-2806
Ring End Gap
  Compression 0.010-0.020
  Oil (Production) 0.010-0.018
  Oil (Service)
    rail 0.015-0.055
    spacer 0.018-0.028
Ring Side Clearance
  Top compression 0.0035-0.0050
  2nd & 3rd
  compression 0.0020-0.0035
  Oil (Production) ...0.0025-0.0040
  Oil (Service) 0.0031-0.0074

Late 806-2806 & 856-2856
Ring End Gap
  Compression 0.010-0.020
  Oil Control 0.010-0.018
Ring Side Clearance
  Top compression 0.0025-0.040
  Second compression 0.0020-0.0035
  Oil control 0.0025-0.0040

Standard cylinder bore is 3.8125-3.8150. Pistons should have a diamet-

ral clearance in cylinder bores of 0.001-0.0045 when measured at bottom of skirt and 90° to piston pin.

PISTONS, SLEEVES AND RINGS

Series 706-2706 Non-Diesel (C263 Engine)

74. The cam ground, aluminum pistons are fitted with two compression rings and one oil control ring and are available in standard size only. New pistons have a diametral clearance of 0.001-0.0045 when measured between piston skirt and installed sleeve at 90° to piston pin.

The dry type cylinder sleeves should be renewed when out-of-round or taper exceeds 0.008. Inside diameter of new sleeve is 3.5593-3.5618. With piston and connecting rod assemblies removed from block, use a suitable puller to remove cylinder sleeves. Clean mating surfaces of block bores and sleeves before new sleeves are installed. After installation, top of sleeves should extend 0.000-0.006 above top surface of block. Sleeve fit in block will vary from 0.0011 interference to 0.0014 loose.

The top compression ring is a chrome barrel face ring and is installed with counterbore up.

The second compression ring is a taper face ring and is installed with largest outside diameter toward bottom of piston. Upper side of ring is marked TOP.

The oil control ring can be installed either side up. A flat spring expander is used with this ring.

Additional piston ring information is as follows:
Ring End Gap
  Compression rings 0.010-0.020
  Oil control ring 0.010-0.018
Ring Side Clearance
  Top compression 0.0025-0.0040
  Second compression ...0.0020-0.0035
  Oil control 0.0025-0.0040

Series 706-2706-756-2756 Non-Diesel (C291 Engine)

75. The cam ground aluminum pistons are fitted with two compression rings and one oil control ring and are available in standard size only. New pistons have a diametral clearance of 0.001-0.0045 when measured between piston skirt and installed sleeve at 90° to piston pin.

The dry type cylinder sleeves should be renewed when out-of-round or taper exceeds 0.008. Inside diameter of new sleeve is 3.7500-3.7525. With piston and connecting rod assemblies removed from block, use a hydraulic sleeve puller to remove cylinder sleeves.

Crankcase bore classifications (1, 2 or 3) are stamped in consecutive order on or near the top of the oil filter mounting pad. Sleeves are available in two outside diameter classes; Class 1 & 2 or Class 2 & 3. A Class 1 & 2 sleeve fits bore classifications 1 or 2 and Class 2 & 3 sleeve fits bore classifications 2 or 3.

Sleeve outside diameters and cylinder block bores are as follows:

Sleeve Type	Sleeve O. D.
Class 1 & 2	3.8765-3.8770
Class 2 & 3	3.8770-3.8775

Cylinder Bore	Bore I. D.
Class 1	3.8750-3.8755
Class 2	3.8755-3.8760
Class 3	3.8760-3.8765

Clean the sleeve and cylinder block bore with solvent and dry with compressed air. Lubricate sleeve and bore with clean diesel fuel. Using a hydraulic sleeve installing tool, press sleeve into cylinder block until sleeve flange is 0.00-0.005 above top surface of cylinder block. The force required to press this sleeve into position will vary from a minimum of 600 pounds to a maximum of 3000 pounds. To translate pounds of force into gage pressure for use with hydraulic ram, the effective area of the ram cylinders must be calculated by the formula d^2 x 0.784. With the ram cylinder effective area calculated, the result can be divided into the stated pounds of force and gage pressure psi range determined. Using an OTC twin piston ram with each piston diameter of 1½ inches, the following example is given: $(d^2 + d^2)$ x 0.784 equals $(2.25 + 2.25)$ x 0.784 equals 3.52 sq. in. effective area of both rams. When 3.52 is divided into 600 (lbs. of force), a gage pressure of 171 psi is obtained. When 3.52 is divided into 3000 (lbs of force), a gage pressure of 852 psi is obtained. Therefore, sleeve installation must occur between 171 and 852 psi gage pressure.

The top compression ring is a chrome barrel face ring and is installed with counterbore up.

The second compression ring is a taper face ring and is installed with largest outside diameter toward bottom of piston. Upper side of ring is marked TOP.

The chrome slotted oil control ring can be installed either side up. A coil spring expander is used with this ring.

Additional piston ring information is as follows:

Ring End Gap
Compression rings 0.010-0.020
Oil control ring 0.010-0.018

Ring Side Clearance
Top compression 0.0025-0.0040
Second compression ..0.0025-0.0035
Oil control 0.0025-0.0040

PISTON PINS

All Non-Diesel Models

76. The full floating type piston pins are retained in the piston bosses by snap rings. Specifications are as follows:

Piston pin diameter 0.8748-0.8749
Diametral clearance
  in piston 0.0002-0.0004
Diametral clearance
  in rod bushing 0.0002-0.0005

Piston pins are available in 0.005 oversize for all non-diesel models.

CONNECTING RODS AND BEARINGS

All Non-Diesel Models

77. Connecting rod bearings are of the slip-in, precision type, renewable from below after removing oil pan and rod caps. When installing new bearing inserts, make certain the projections on same engage slots in connecting rod and cap and that cylinder identifying numbers on rod and cap are in register and face toward camshaft side of engine. Connecting rod bearings are available in standard size and undersizes of 0.002, 0.010, 0.020 and 0.030. Check the crankshaft crankpins and connecting rod bearings against the values which follow:

Crankpin diameter 2.373-2.374
Max. allowable out-of-round ..0.0015
Max. allowable taper 0.0015
Rod bearing diametral
  clearance 0.009-0.0034
Rod side clearance 0.007-0.013
Rod bolt torque
  PLACE bolt* ....50 ft.-lbs.
  PITCH bolt* ....45 ft.-lbs.
  *Refer to paragraph 72 for bolt identification.

CRANKSHAFT AND MAIN BEARINGS

All Non-Diesel Models

78. Crankshaft is supported in four main bearings and thrust is taken by the third (rear intermediate) bearing. Main bearings are of the shimless, non-adjustable, slip-in precision type, renewable from below after removing the oil pan and main bearing caps. Removal of crankshaft requires R&R of engine. Check crankshaft and main bearings against the values which follow:

Crankpin diameter 2.373-2.374
Main journal diameter 2.748-2.749

Max. allowable out-of-round ..0.0015
Max. allowable taper 0.0015
Crankshaft end play 0.005-0.013
Main bearing diametral
  clearance 0.0012-0.0042
Main bearing bolt torque ..80 ft.-lbs.

Main bearings are available in standard size and undersizes of 0.002, 0.010, 0.020 and 0.030. Alignment dowels (IH tool FES 6-1 or equivalent) should be used when installing the rear main bearing cap.

CRANKSHAFT SEALS

All Non-Diesel Models

79. FRONT. To renew the crankshaft front oil seal, first remove hood, drain cooling system and disconnect radiator hoses. On Farmall models, identify and disconnect power steering lines. On International models, disconnect forward end of power steering cylinder from center steering arm. Then, on all models, identify and disconnect the hydraulic oil cooler lines. Support tractor under clutch housing. On Farmall models equipped with wide front axle and all International models, disconnect stay rod ball. On all models, attach hoist to front support, unbolt front support from side rails and roll complete front assembly from tractor. Remove fan and generator or alternator drive belts. Remove crankshaft pulley retaining nut, attach a suitable puller and remove crankshaft pulley. Remove and renew oil seal in conventional manner. Drive new seal in until it is seated against shoulder in timing gear cover. Install crankshaft pulley and tighten retaining nut to a torque of 95 ft.-lbs. Reassemble tractor by reversing disassembly procedure.

80. REAR. To renew the crankshaft rear oil seal, the engine must be detached from clutch housing as outlined in paragraph 229. Then, unbolt and remove clutch and flywheel. The lip type seal can be removed after collapsing same. Take care not to damage sealing surface of crankshaft when removing seal. Use seal installing tool and oil seal driver (IH tools FES6-2 and FES6-3) or equivalent and drive seal in until it is flush with rear of crankcase. Lip of seal must be toward front of engine.

FLYWHEEL

All Non-Diesel Models

81. To remove the flywheel, first split tractor as outlined in paragraph 229, then unbolt and remove clutch assembly. Remove six cap screws and lift flywheel from crankshaft. When

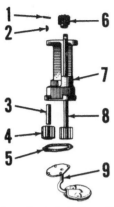

Fig. IH155 — Exploded view of oil pump used on non-diesel engines.

1. Pin
2. Woodruff key
3. Idler gear shaft
4. Idler gear
5. Gasket
6. Drive gear
7. Pump body
8. Drive shaft and gear
9. Cover and screen

installing flywheel, coat cap screws with sealer and tighten cap screws to a torque of 75 ft.-lbs.

To install a new flywheel ring gear, heat same to approximately 500 degrees F.

OIL PUMP

All Non-Diesel Models

82. The gear type oil pump is gear driven from a pinion on camshaft and is accessible for removal after removing the engine oil pan. Disassembly and overhaul of pump is obvious after an examination of the unit and reference to Fig. IH155. Gaskets (5) between pump cover and body can be varied to obtain the recommended 0.0025-0.0055 pumping (body) gear end play.
Refer to the following specifications:
Pumping gears recommended
  backlash 0.003-0.006
Pump drive gear
  recommended backlash ..0.000-0.008
Pumping gear end play .0.0025-0.0055
Gear teeth to body
  radial clearance 0.0068-0.0108
Drive shaft clearance ...0.0015-0.0030
Mounting bolt torque ......22 ft.-lbs.

Service (replacement) pump shaft and gear assemblies are not drilled to accept the pump driving gear pin. A

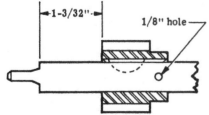

Fig. IH156—New pump shaft and gear assemblies will require a ⅛-inch hole to be drilled at location shown.

⅛-inch hole must be drilled through the shaft after the gear is installed on the shaft to the dimension shown in Fig. IH156.

NOTE: When installing the oil pump on engine, time the pump as follows: Crank engine until number one piston is coming up on compression stroke. Continue cranking until the TDC mark on crankshaft pulley or flywheel is in register with the timing pointer. Install oil pump so that tang on oil pump shaft is in the approximate position shown in Fig. IH157. Retime distributor as outlined in paragraph 212 or 214.

OIL PRESSURE RELIEF VALVE

All Non-Diesel Models

83. Series 706 and 2706 non-diesel engines prior to serial number 47104 and series 806 and 2806 non-diesel engines prior to serial number 5146 are equipped with the cast iron base oil filter assembly shown in Fig. IH158. The spring loaded, plunger type pressure regulator (relief) valve is located in the filter base and is non-adjustable. Spring (13) should be installed with the closed coils in the plunger. Relief valve piston (12) seals on outer diameter of valve bore. Inner end of bore is a stop for the piston and is not the valve seat. If pressure is lower than normal, check spring tension and inspect valve and bore for excessive wear, scoring or other damage. Element by-pass valve is located in center tube (1). By-pass valve (3) and spring (4) can be removed from center tube after driving out pin (2). Specifications are as follows:

Pressure regulating valve,
  Valve diameter 0.743-0.745
  Clearance in bore 0.002-0.007
  Spring free length ....3.0 in.
  Spring test and
    length ......18 lbs @ 1 13/16 in.
By-pass valve,
  Spring free length ......2 15/64 in.
  Spring test and
    length ....3.4 lbs. @ 2.0 in.
Oil pressure at
  1800 RPM ....30-40 psi

84. Series 706 and 2706 non-diesel engines with serial number 47104 and above, series 806 and 2806 non-diesel engines with serial number 5146 and above and all non-diesel series 756, 2756, 856 and 2856 are equipped with the die cast base oil filter assembly shown in Fig. IH159. The pressure regulator (relief) valve, by-pass valve and check valve are located in the oil filter base (12). All valves and their springs are retained in position by snap rings and removal is obvious after an examination of the unit and

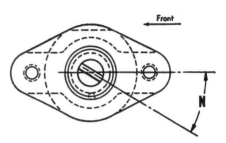

Fig. IH157—On non-diesel engines, position No. 1 piston at TDC on compression stroke and mesh oil pump drive gear so angle (N) of drive shaft tang is approximately 30° to centerline of engine.

reference to Fig. IH159. Specifications are as follows:
Check valve,
  Valve diameter 0.770
  Spring free length ....1 5/64 in.
By-pass valve,
  Valve diameter 0.770
  Spring free length .... 2⅞ in.
  Spring test and
    length ......4.5 lbs. @ 1 5/64 in.
Oil pressure at 1800 RPM ..30-40 psi

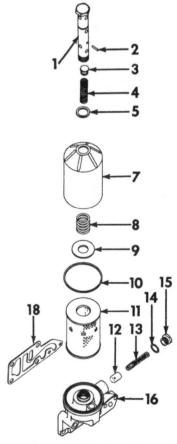

Fig. IH158—Exploded view of cast iron base oil filter assembly used on early production series 706, 2706, 806 and 2806 non-diesel engines. Refer to text.

1. Center tube
2. Pin
3. By-pass valve
4. Spring
5. Gasket
7. Case
8. Hold-down spring
9. Retainer
10. Gasket
11. Element
12. Relief valve piston
13. Spring
14. Gasket
15. Cap
16. Filter base
18. Gasket

OIL PAN

All Non-Diesel Models

85. Removal of oil pan is conventional and on tricycle model tractors, can be accomplished with no other disassembly.

On Farmall models with adjustable wide front axle, the stay rod bracket must be removed from side rails before oil pan can be removed.

On International and "All Wheel Drive" models, disconnect stay rod (or bracket) from clutch housing and front axle from front support and raise front of tractor to provide clearance for oil pan removal.

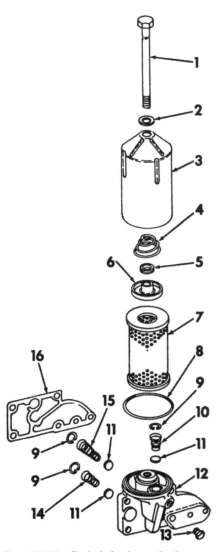

Fig. IH159—Exploded view of die cast base oil filter assembly used on late production 706, 2706, 806 and 2806 and all 756, 2756, 856 and 2856 non-diesel engines.

1. Center tube
2. Gasket
3. Case
4. Hold-down spring
5. Grommet
6. Retainer
7. Element
8. Gasket
9. Snap ring
10. Check valve spring
11. Valve
12. Filter base
13. Drain plug
14. Relief valve spring
15. By-pass spring
16. Gasket

ENGINE AND COMPONENTS (DIESEL)

Farmall 706 diesel tractors prior to serial number 37237 and International 706 and 2706 diesel tractors prior to serial number 5274 are equipped with engines having a bore and stroke of 3 11/16 x 4 25/64 inches and a displacement of 282 cubic inches. Later series 706 and 2706 and all series 756 and 2756 diesel tractors are equipped with engines having a bore and stroke of 3⅞ x 4⅜ inches and a displacement of 310 cubic inches.

Series 806, 2806, 1206 and 21206 diesel tractors are equipped with engines having a bore and stroke of 4⅛ x 4½ inches and a displacement of 361 cubic inches.

Series 856, 2856, 1256, 21256, 1456 and 21456 diesel tractors are equipped with engines having a bore and stroke of 4 5/16 x 4⅝ inches and a displacement of 407 cubic inches.

All engines are a six cylinder design. The 310 cubic inch engine is equipped with wet type sleeves and all other engines are fitted with dry type sleeves. Dry type air filters are used in the air induction systems of all engines. Series 1206, 21206, 1256, 21256, 1456 and 21456 diesel engines are equipped with turbochargers.

R&R ENGINE WITH CLUTCH

All Diesel Models

86. Removal of engine is best accomplished by removing the front support radiator, front axle and wheels assembly and the fuel tank assembly, then unbolting and removing the engine and side rails from the clutch housing as follows.

Drain cooling system, disconnect battery cables and remove hood. Disconnect radiator hoses and air cleaner inlet hose. Identify and disconnect power steering lines and hydraulic oil cooler lines. Remove operator assist handles, torque amplifier lever and notice that TA lever and shaft have match marks affixed to insure correct reassembly. Remove center hood and steering support cover. Disconnect all interfering lines and wiring from fuel tank and support. Attach hoist to tank, unbolt and lift fuel tank and support from tractor.

Disconnect remaining engine controls and electrical wiring. Support tractor under clutch housing, attach hoist to front support, unbolt front support from side rails, disconnect stay rod ball, if so equipped, and roll complete front assembly from tractor.

NOTE: If additional clearance is needed for removal of front support from side rails, loosen the engine front mounting cap screws.

Attach hoist to engine, then unbolt and remove engine and side rails from clutch housing.

When reinstalling engine assembly, unbolt and remove clutch assembly from flywheel. Place pressure plate assembly and drive disc on shafts in clutch housing. Clutch can be bolted to flywheel after engine is installed by working through the opening at bottom of clutch housing.

CYLINDER HEAD

All Diesel Models

87. To remove the cylinder head, first remove fuel tank assembly as outlined in paragraph 86. Drain cooling system and remove coolant temperature bulb. Unbolt and remove air cleaner assembly and on series 1206, 21206, 1256, 21256, 1456, and 21456, remove turbocharger as outlined in paragraph 191.

On series 706 and 2706 equipped with D282 engine, unbolt and remove inlet and exhaust manifolds. Disconnect glow plug wires at junction block, unbolt fuel filter bracket from cylinder head and allow filters to rest on frame channel. Disconnect injector lines from fuel injectors and injection pump. NOTE: Cap all fuel connections immediately to prevent entrance of dirt or other foreign material. Remove rocker arm cover, rocker arms and shaft assembly, and push rods. Remove cylinder head cap screws and lift off cylinder head.

On series 706, 2706, 756 and 2756 equipped with D310 engine, disconnect injection lines from fuel injectors and injection pump and remove excess fuel return line from injectors. Cap all fuel connections immediately. CAUTION: Injector nozzle assemblies protrude slightly through combustion side of cylinder head. It is recommended that injector nozzles be removed before removing cylinder head. Unbolt and remove exhaust manifold and water collecting tube (manifold) from right side and inlet manifold from left side of cylinder head. Remove rocker arm cover, rocker arms and shaft assembly and push rods. Remove cylinder head retaining nuts and lift off cylinder head.

On series 806, 2806, 856, 2856, 1206, 21206, 1256, 21256, 1456 and 21456 (361 and 407 cubic inch engines), remove the breather outlet tube from left front of cylinder head. Disconnect the external oil line from left side of cylinder head. Disconnect elbows of the oil filter tubes from engine oil cooler and pressure regulator block, then unbolt and remove filters from cylinder head. Loosen by-pass hose clamps and remove thermostat housing. Unbolt and remove heat shields and exhaust manifold. Allow fuel filters to rest on frame channel. Disconnect injection lines from fuel injectors and injection pump. Remove excess fuel return line from injectors. Cap all fuel connections immediately. CAUTION: Injector nozzle assemblies protrude slightly (0.087-0.120) through the combustion side of cylinder head. It is recommended that injector nozzles be removed before removing cylinder head. Remove rocker arm cover, rocker arms and shaft assembly and push rods. Remove cylinder head retaining cap screws and remove cylinder head.

On all models, check cylinder head for warpage as follows: Place a straight edge across the machined (combustion) side of cylinder head and measure between cylinder head and straight edge. If a 0.003 feeler gage can be inserted between straight edge and cylinder head within any six-inch distance, the cylinder head should be refaced with a surface grinder; PROVIDING, not more than 0.010 of material is removed. The standard distance from rocker arm cover surface to combustion surface of cylinder head is as follows:

D282 ....4.015-4.025
D310 ....3.890-3.910
D361, DT361,
  D401 & DT407 ....5.098-5.102

NOTE: After cylinder head is resurfaced, check the valve head recession or protrusion specifications outlined in paragraphs 89, 90 and 91. Correct as necessary.

When reinstalling cylinder head on all models, use new head gasket and make certain that gasket sealing surfaces are clean and dry. DO NOT use sealants or lubricants on head gasket, cylinder head or block. Guide studs should be used when installing cylinder head on D282 engines.

CAUTION: Because of the minimum amount of clearance that exists between valves and piston tops, loosen the rocker arm adjusting screws before installing the rocker arms and shaft assembly. Refer to paragraphs 89, 90 and 91 for information concerning valve adjustment.

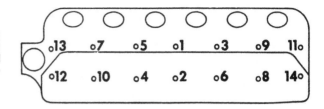

Fig. IH160 — Series 706 and 2706 diesel (D282 engine) cylinder head cap screw tightening sequence.

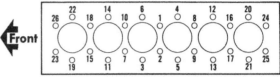

Fig. IH161—Series 706, 2706, 756 and 2756 diesel (D310 engine) cylinder head retaining nuts tightening sequence.

1st Torque	50 FT.-LBS.
2nd Torque	70 FT.-LBS.
3rd Torque	90 FT.-LBS.

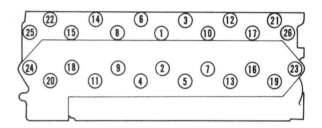

Fig. IH162 — Cylinder head cap screw tightening sequence for all diesel series equipped with D-361, DT361, D407 and DT407 engines.

On series 706 and 2706 equipped with D282 engines, tighten cylinder head cap screws in two steps using the sequence shown in Fig. IH160. Torque the cap screws to 60-70 ft.-lbs. during the first step and to 110-120 ft.-lbs. during the second step.

On series 706, 2706, 756 and 2756 equipped with D310 engines, tighten cylinder head retaining nuts in three steps using the sequence shown in Fig. IH161. Tighten the nuts to a torque of 50 ft.-lbs. during the first step, 70 ft.-lbs. during the second step and 90 ft.-lbs. during the third step.

On series 806, 2806, 856, 2856, 1206, 21206, 1256, 21256, 1456, and 21456 equipped with D361, DT361, D407 and DT407 engines, tighten cylinder head cap screws in three steps using the sequence shown in Fig. IH162. Tighten the cap screws to a torque of 65 ft.-lbs. during the first step, 110 ft.-lbs. during the second step and 135 ft.-lbs. during the third step.

COOLING TUBES AND NOZZLE SLEEVES

All Diesel Engines Except D282

88. The cylinder head is fitted with brass injector nozzle sleeves which pass through the coolant passages. In addition, cooling jet tubes (one for each cylinder) are used to direct a portion of the coolant to the valve seat and nozzle sleeve area. Both the nozzle sleeves and cooling tubes are available as service items.

To renew the nozzle sleeves, remove injectors and cylinder head as outlined in paragraph 87. On D310 engines, use special bolt (IH tool No. FES 112-4) and turn it into the sleeve. Attach a slide hammer puller and remove the sleeve. See Fig. IH163. On D361, DT-361, D407 and DT407 engines, use adapter (IH tool No. FES 25-9) and

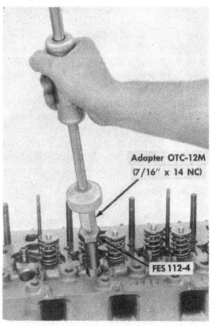

Fig. IH163—On D310 engines, use IH tool No. FES 112-4 and slide hammer to remove injector nozzle sleeves.

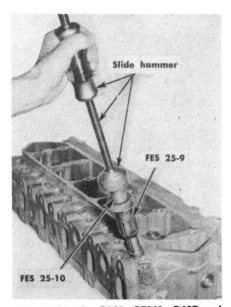

Fig. IH164—On D361, DT361, D407 and DT407 engines, use special IH tools shown to remove injector nozzle sleeves.

Fig. IH166—Drive nozzle sleeves into D310 cylinder head until they bottom.

expanding screw (IH tool FES 25-10) along with a slide hammer and pull nozzle sleeve as shown in Fig. IH164.

NOTE: Use caution during sleeve removal not to damage the sealing areas in the cylinder head. Under no circumstances should screwdrivers, chisels or other such tools be used in an attempt to remove injector nozzle sleeves.

When installing nozzle sleeves, be sure the sealing areas are completely clean and free of scratches. Apply a light coat of "Grade B Loctite" on sealing surfaces of nozzle sleeve. On D310 engines, use installing tool (IH tool No. FES 112-3) shown in Fig. IH165 and drive injector nozzle sleeves into their bores until they bottom as shown in Fig. IH166. On D361, DT361, D407 and DT407 engines use installing tool (IH tool FES 68-5) shown in Fig. IH167 and install nozzle sleeves using same procedure.

NOTE: Injector nozzle sleeves have an interference fit in their bores. When installing the sleeves, be sure sleeve is driven straight with its bore and is completely bottomed.

To remove the cooling jet tubes, thread the inside diameter and install

a cap screw to assist in removal. Install new cooling tubes so coolant is directed between the valves of each cylinder. Tubes must be installed flush with cylinder head surface.

VALVES AND SEATS

Series 706-2706 (D282 Engine)

89. Inlet and exhaust valves are not interchangeable. Both the inlet and exhaust valves seat directly in the cylinder head. Valve rotators (Rotocoils) are used on both inlet and exhaust valves. Deflector type stem seals are used on all valves.

Check the valves and seats against the following specifications:

Inlet and Exhaust

Face and seat angle ....45°
Stem diameter 0.3715-0.3725
Stem to guide diametral
  clearance 0.0015-0.004
Seat width 0.066-0.080
Valve run-out (max.) 0.002
Valve tappet gap (warm) 0.027
Valve recession from face of
  cylinder head 0.003-0.046

To adjust valve tappet gap, crank engine to position number one piston at top dead center of compression stroke. Adjust the six valves indicated on the chart shown in Fig. IH168.

Turn engine crankshaft one complete revolution to position number six piston at TDC (compression) and adjust the remaining six valves indicated on chart.

Series 706-2706-756-2756 (D310 Engine)

90. Inlet and exhaust valves are not interchangeable. The inlet valves seat directly in the cylinder head and the exhaust valves seat on renewable seat inserts. Inserts are available in oversizes of 0.004 and 0.016. Valve rotators (Rotocaps) and valve stem seals are used on all valves.

When removing valve seat inserts, use the proper puller. Do not attempt to drive chisel under insert as counterbore will be damaged. Chill new insert with dry ice or liquid Freon prior to installing. When new insert is properly bottomed it should be 0.008-0.030 below edge of counterbore. After installation, peen the cylinder head material around the complete outer circumference of the valve seat insert.

Check the valves and seats against the following specifications:

Inlet

Face and seat angle ....45°
Stem diameter 0.3919-0.3923

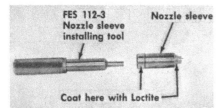

Fig. IH165—Injector nozzle sleeve installing tool used on D310 engines. Apply "Grade B Loctite" to sealing surfaces of nozzle sleeve.

Fig. IH167 — On D361, DT361, D407 and DT407 engines, use installing tool shown and drive nozzle sleeves in cylinder head bores until they bottom.

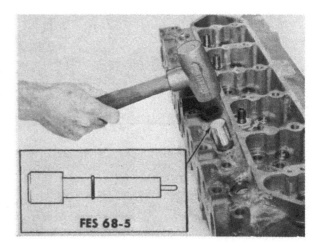

	ADJUST VALVES (ENGINE WARM)											
With No. 1 Piston at T.D.C. (Compression)	1	2	3		5		7		9			
With No. 6 Piston at T.D.C. (Compression)				4		6		8		10	11	12

← Front   Rear →

1 2 3 4 5 6 7 8 9 10 11 12

Numbering sequence of valves which correspond to chart

Fig. IH168—Chart shows valve tappet gap adjusting procedure used on series 706 and 2706 diesel (D282) engines. Refer to text.

Stem to guide diametral clearance,
Normal 0.0014-0.0026
Maximum allowable 0.006
Seat width 0.076-0.080
Valve run-out (max.) 0.001
Valve tappet gap (warm) 0.010
Valve recession from face of cylinder head,
Normal 0.039-0.055
Maximum allowable 0.120

Exhaust
Face and seat angle .... 45°
Stem diameter 0.3911-0.3915
Stem to guide diametral clearance,
Normal 0.0022-0.0034
Maximum allowable 0.006
Seat width 0.081-0.089
Valve run-out (max.) 0.001
Valve tappet gap (warm) 0.012
Valve recession from face of cylinder head,
Normal 0.047-0.063
Maximum allowable 0.120

CAUTION: Due to close clearance between valves and pistons, severe damage can result from inserting feeler gage between valve stem and valve lever (rocker arm) with engine running. DO NOT attempt to adjust valve tappet gap with engine running.

To adjust valve tappet gap, crank engine to position number one piston at top dead center of compression stroke. Adjust the six valves indicated on the chart shown in Fig. IH169.

NOTE: The valve arrangement is exhaust-intake-exhaust-intake and so on, starting from front of cylinder head.

Turn engine crankshaft one complete revolution to position number six piston at TDC (compression) and adjust the remaining six valves indicated on chart.

Series 806-2806-856-2856-1206-21206-1256-21256-1456-21456 Diesel

91. Inlet and exhaust valves are not interchangeable. The inlet valves seat directly in the cylinder head and the exhaust valves seat on renewable seat inserts. However, inlet valve seat inserts are available for service. Both inlet and exhaust valve seat inserts are available in standard size and oversizes of 0.005 and 0.015. Valve rotators (Rotocoils) are used on the exhaust valves.

When removing valve seats, use the proper puller. Do not attempt to drive chisel under insert as counterbore will be damaged. Chill new insert with dry ice or liquid Freon prior to installing. When new insert is properly bottomed, it should be 0.008-0.030 below edge of counterbore. After installation, peen the cylinder head material around the complete outer circumference of the valve seat insert.

Check the valves and seats against the specifications which follow:

Inlet and Exhaust
Face and seat angle .... 45°
Stem diameter 0.4348-0.4355
Stem to guide diametral clearance,
Normal 0.0015-0.0032
Maximum allowable 0.008
Seat width,
Inlet 5/64-inch
Exhaust 3/32-inch
Valve run-out (max.) 0.001
Valve tappet gap (warm),
Inlet 0.013
Exhaust 0.025

With valves installed in cylinder head, measure valve head position in relation to cylinder head surface. The inlet valve head should be flush plus

or minus 0.0065 with cylinder head surface. Maximum allowable recession is 0.0225. The exhaust valve head should protrude 0.0815-0.0945 above cylinder head surface. Minimum allowable protrusion is 0.0655.

CAUTION: Due to the close valve to piston clearance, severe damage can result from inserting feeler gage between valve stem and valve lever (rocker arm) with engine running. DO NOT attempt to adjust valve tappet gap with engine running.

To adjust valve tappet gap, proceed as follows: Remove timing hole cover on right side of flywheel housing. Using a pry bar on flywheel teeth, turn flywheel in normal direction of rotation to position number one piston at TDC of compression stroke. Adjust the six valves indicated on the chart shown in Fig. IH169.

NOTE: The valve arrangement is exhaust-intake-exhaust-intake and so on, starting from front of cylinder head.

Turn flywheel one complete revolution in normal direction of rotation to position number six piston at TDC (compression) and adjust the remaining six valves indicated on chart.

VALVE GUIDES AND SPRINGS
Series 706-2706 (D282 Engine)

92. The inlet and exhaust valve guides are interchangeable. Inlet and exhaust guides should be pressed in cylinder head until top of guides is 15/16-inch above spring recess in the head. Guides are pre-sized and, if carefully installed, should need no final sizing. Inside diameter should be 0.3740-0.3755 and valve stem to guide diametral clearance should be 0.0015-0.004.

Inlet and exhaust valve springs are also interchangeable. Springs should have a free length of 2¼ inches and should test 159-164 pounds when compressed to a length of 1.359 inches. Renew any spring which is rusted, discolored or does not meet the pressure test specifications.

Series 706-2706-756-2756 (D310 Engine)

93. The inlet and exhaust valve guides should be pressed into cylinder head until top of guides is 1 5/32 inches above spring recess in head. After installation, guides must be reamed to an inside diameter of 0.3940-0.3945. Valve stem to guide diametral clearance should be 0.0014-0.0026 for inlet valves and 0.0022-0.0034 for exhaust valves. Maximum allowable stem clearance in all guides is 0.006.

	ADJUST VALVES (ENGINE WARM)											
With No. 1 Piston at T.D.C. (Compression)	1	2		4	5			8	9			
With No. 6 Piston at T.D.C. (Compression)			3			6	7			10	11	12

← Front

1 2 3 4 5 6 7 8 9 10 11 12

Numbering sequence of valves which correspond to chart

Fig. IH169—Chart shows valve tappet gap adjusting procedure used on D310, D361, DT361, D407 and DT407 engines. Refer to text.

Inlet and exhaust valve springs are also interchangeable. Springs should have a free length of 2.173 inches and should test 146-160 pounds when compressed to a length of 1 11/32 inches. Renew any spring which is rusted, discolored or does not meet the pressure test specifications.

Series 806-2806-856-2856-1206-21206-1256-21256-1456-21456 Diesel

94. The inlet and exhaust valve guides are interchangeable in the D361, DT361, D407 and DT407 engines. Inlet and exhaust valve guides should be pressed into cylinder head until top of guides is 1 1/16 inches above spring recess in head. The inside diameter of valve guides is knurled for oil control. Guides are pre-sized; however, since they are a press fit in cylinder head, it is necessary to ream them to remove any burrs or slight distortion caused by the pressing operation. Inside diameter should be 0.0015-0.0032 with maximum allowable clearance of 0.008.

Each valve is equipped with two (inner and outer) valve springs and springs are not interchangeable between inlet and exhaust valves. Valve spring specifications are as follows:

Inlet valve springs
Free length,
Inner . . . . .2 11/32 in.
Outer . . . . .2 9/16 in.
Test load and length,
Inner . . . . .85 lbs. @ 1½ in.
Outer . . . . . . .136 lbs. @ 1 45/64 in.
Exhaust valve springs
Free length,
Inner . . . . 1 63/64 in...
Outer . . . . 2 21/64 in.
Test load and length,
Inner . . . . . . .81 lbs. @ 1 7/32 in.
Outer . . . . . . .135 lbs. @ 1 7/16 in.

Valve rotators (Rotocoil) are used under exhaust valve springs and spring spacers are used under inlet valve springs. Install springs with dampener coils towards cylinder head. Renew any spring which is rusted, discolored or does not meet test specifications.

VALVE TAPPETS (CAM FOLLOWERS)

Series 706-2706 (D282 Engine)

95. Valve tappets are the barrel type and operate in unbushed bores in crankcase. Tappet diameter is 0.9965-0.9970 and should operate with a clearance of 0.002-0.004 in the 0.9990-1.0005 crankcase bores. Oversize tappets are not available. Tappets can be removed from side of crankcase after

removing rocker arm cover, rocker arms, push rods and side cover plate.

Series 706-2706-756-2756 (D310 Engine)

96. The 0.7862-0.7868 diameter mushroom type tappets operate directly in the unbushed crankcase bores. Clearance of tappets in the bores should be 0.0005-0.0024. Tappets can be removed after removing the oil pan and the camshaft as outlined in paragraph 124. Oversize tappets are not available.

Series 806-2806-856-2856-1206-21206-1256-21256-1456-21456 Diesel

97. Valve tappets used in D361, DT-361, D407 and DT407 engines are the mushroom type and operate in unbushed bores in crankcase. Tappet diameter is 0.623-0.624 and should operate with a clearance of 0.001-0.004 in the 0.625-0.627 crankcase bores. Tappets are available in standard size only and removal requires the removal of camshaft as outlined in paragraph 126 and engine oil pan.

VALVE ROCKER ARM COVER

All Diesel Models

98. Removal of the rocker arm cover is obvious after removal of front hood. Disconnect air inlet from inlet manifold and on models so equipped, remove fuel tank left front support. On models equipped with D282 engine, remove inlet manifold.

When reinstalling cover, use new gasket to insure an oil tight seal.

VALVE TAPPET LEVERS (ROCKER ARMS)

Series 706-2706 (D282 Engine)

99. **REMOVE AND REINSTALL.** Removal of the rocker arms and shaft assembly may be accomplished in some cases without raising the fuel tank and bracket assembly; however, the rocker shaft bracket hold down screws are also the left row of cyl-

inder head cap screws. Removal of these screws may allow the left side of cylinder head to raise slightly and damage to the head gasket could result. Because of the possible damage to the head gasket, it is recommended that the cylinder head be removed as in paragraph 87 when removing the rocker arm assembly.

100. **OVERHAUL.** The rocker arm shaft is two-piece and has an outside diameter of 0.748-0.749. Rocker arm bushings are not renewable. Inside diameter of bushings is 0.7505-0.752. Inlet valve rocker arms are interchangeable; but exhaust valve rocker arms are offset towards the nearest rocker shaft bracket. See Fig. IH170. Rocker arms may be refaced providing the original contour is maintained.

Number 2, 4 and 6 rocker shaft brackets are equipped with dowel sleeves for alignment.

Series 706-2706-756-2756 (D310 Engine)

101. **REMOVE AND REINSTALL.** Removal of the rocker arms and shaft assembly is conventional after removal of rocker arm cover. When reinstalling, tighten the hold-down nuts on bracket studs to a torque of 47 ft.-lbs.

102. **OVERHAUL.** To remove the rocker arms from the one-piece shaft, remove bracket clamp bolts and slide all parts from shaft. Outside diameter of rocker shaft is 0.8491-0.8501. The renewable bushings in rocker arms should have an operating clearance of 0.0009-0.0025 on rocker shaft with a maximum allowable clearance of 0.008. Rocker arms may be refaced providing the original contour is maintained. Rocker arm adjusting screws are self-locking. If they turn with less than 12 ft.-lbs. torque, renew adjusting screw and/or rocker arm.

Reassemble rocker arms on rocker shaft keeping the following points in mind. Thrust washers are used between each spring and rocker arms and spacer rings are used between rocker arms and brackets except between rear rocker arm and rear

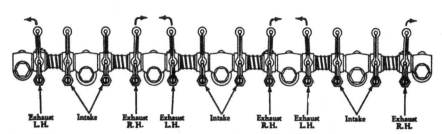

Fig. IH170—On Series 706 and 2706 equipped with D282 engines, inlet valve rocker arms are interchangeable but exhaust valve rocker arms are left and right hand units and are offset towards the nearest shaft backet as shown.

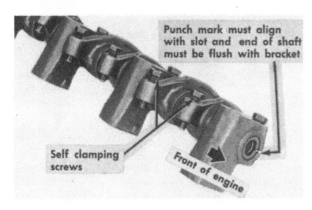

Punch mark must align with slot and end of shaft must be flush with bracket

Self clamping screws

Front of engine

Fig. IH171 — On series 706, 2706, 756 and 2756 equipped with D310 engines, align punch mark on front end of rocker shaft with slot in front mounting bracket. End of shaft must be flush with bracket.

bracket. To insure that lubrication holes in rocker shaft are in correct position, align punch mark on front end of shaft with slot in front mounting bracket as shown in Fig. IH171. End of shaft must also be flush with bracket. Rocker shaft clamp screws on brackets should be tightened to a torque of 10 ft.-lbs.

Series 806-2806-856-2856-1206-21206-1256-21256-1456-21456 Diesel

103. **REMOVE AND REINSTALL.** Removal of rocker arms and shaft assembly is accomplished as follows: Remove front hood, center hood and steering support cover. Remove air inlet elbow and hose from cylinder head inlet manifold. Disconnect fuel tank supports from engine and completely remove left front tank support. Remove rear tank support bolts except the top one, then unclip and/or disconnect all interfering wires, cables and piping. Pivot front of fuel tank upward and block in position. Rocker arm cover and the rocker arms and shaft assembly can now be removed.

104. **OVERHAUL.** The rocker shaft is a one-piece unit with an outside diameter of 0.872-0.873. Inside diameter of rocker arm bushings is 0.8745-0.8760 which provides and operating clearance of 0.0015-0.004. Maximum allowable clearance is 0.007. All rocker arms are interchangeable and bushings are renewable in rocker arms. Rocker arms may be refaced providing the original contour is maintained. Rocker shaft ends are fitted with plugs and numbers 1 and 5 shaft brackets have locating pins. Rocker arms, springs and brackets can be removed from shaft after removing retainers from ends of shaft.

VALVE ROTATORS

All Diesel Models

105. Two types of positive valve rotators are used (Rotocaps and Roto-

coils). Rotocaps are used on the D310 engine and Rotocoils are used on all other engines.

Normal servicing of the valve rotators consists of renewing the units. It is important to observe the valve action after engine is started. Valve rotator action can be considered satisfactory if the valve rotates a slight amount each time the valve opens.

The Rotocoil rotates the valve at a slower speed than the Rotocap. Rotocaps should not be installed on engines where Rotocoil rotators are specified.

VALVE TIMING

Valve timing is correct when the timing (punch) marks on the timing gear train are properly aligned as shown in Figs. IH172, IH173 or IH174. To check valve timing on an assembled engine, follow the procedure outlined in the following paragraphs.

Series 706-2706 (D282 Engine)

106. To check valve timing, remove rocker arm cover and crank engine to position number one piston at TDC of compression stroke. Adjust number one cylinder intake valve tappet gap to 0.030. Slowly turn crankshaft in normal direction of rotation until valve lever becomes tight against number one intake valve stem and the push rod can no longer be turned by hand. At this point, number one intake valve will start to open and timing pointer should be at or near the 12 degree BTDC mark on crankshaft pulley or flywheel.

NOTE: One tooth "out of time" equals approximately 13 degrees.

Readjust number one intake valve tappet gap as outlined in paragraph 89.

Series 706-2706-756-2756 (D310 Engine)

107. To check valve timing, remove rocker arm cover and crank engine to position number one piston at TDC of compression stroke. Adjust number one cylinder intake valve tappet gap

to 0.014. Place a 0.004 feeler gage between valve lever and valve stem of number one intake valve. Slowly rotate crankshaft in normal direction until valve lever becomes tight on feeler gage. At this point, number one intake valve will start to open and timing pointer should be within the range of 23 to 29 degrees BTDC on the flywheel.

NOTE: One tooth "out of time" equals approximately 11 degrees.

Readjust number one intake valve tappet gap as outlined in paragraph 90.

Series 806-2806-856-2856-1206-21206-1256-21256-1456-21456 Diesel

108. To check valve timing, remove rocker arm cover and crank engine to position number one piston at TDC of compression stroke. Adjust number one cylinder intake valve tappet gap to 0.017. Place a 0.004 feeler gage between valve lever and valve stem of number one intake valve. Slowly rotate crankshaft in normal direction until valve lever becomes tight on feeler gage. At this point, number one intake valve will start to open and timing pointer should be within the range of 22 to 28 degrees BTDC on the flywheel.

NOTE: One tooth "out of time" equals approximately 12 degrees.

Readjust number one intake valve tappet gap as outlined in paragraph 91.

TIMING GEAR COVER

All Diesel Models

109. To remove the timing gear cover, first remove hood, drain cooling system and disconnect battery cables. Disconnect air cleaner inlet hose and radiator hoses. On International models, disconnect forward end of power steering cylinder from center steering arm. On Farmall models, identify and disconnect power steering lines. On all models, identify and disconnect hydraulic oil cooler lines. Plug or cap openings to prevent dirt or other foreign material from entering hydraulic system. Support tractor under clutch housing. On Farmall models equipped with wide front axle and all International models, disconnect stay rod ball. Attach hoist to front support, unbolt front support from side rails and roll complete front assembly forward from tractor Unbolt and remove fan, generator or alternator and drive belts.

On series 706 and 2706 equipped with D282 engines, remove crankshaft nut and clamp a bearing splitter deep in the fan drive belt groove of the crankshaft pulley. Attach a puller to bearing splitter and remove crankshaft pulley. Remove cap screws retaining oil pan to timing gear cover and loosen the remaining oil pan cap screws. Unbolt timing gear cover, then pull cover forward off the dowels and remove from engine. Use care not to damage oil pan gasket. Reassemble by reversing the disassembly procedure. Tighten timing gear cover retaining cap screws to the following torques; ½-inch capscrews, 90 ft.-lbs; ⅜-inch cap screws, 38 ft.-lbs. Crankshaft pulley retaining nut should be tightened to a torque of 95 ft.-lbs.

On series 706, 2706, 756 and 2756 equipped with D310 engines, unbolt and remove air cleaner assembly. Remove pump pulley, then unbolt and remove water pump and carrier assembly. Disconnect tachometer drive cable and remove tachometer drive unit from front cover. Do not lose the small driving tang when removing tachometer drive. Remove the three cap screws, flat washer and pressure ring from crankshaft. Tap crankshaft pulley with a plastic hammer to loosen pulley, then slide pulley off of the wedge rings. Remove cap screws retaining oil pan to timing gear cover and loosen the remaining oil pan cap screws. Unbolt and remove timing gear cover. Reassemble by reversing the disassembly procedure. When installing the crankshaft pulley, place one pressure ring on crankshaft with thick end towards engine. Install the wedge rings in pulley bore so that slots in rings are 90 degrees apart. Slide pulley onto crankshaft and align with timing pin. Install pressure ring, flat pressure washer and three cap screws. Tighten the cap screws evenly to a torque of 55 ft.-lbs.

Fig. IH172—Gear train and timing marks on series 706 and 2706 equipped with D282 engines. Note double row of teeth on idler gear.

CA. Camshaft gear
CR. Crankshaft gear
ID. Idler gear
IN. Injection pump drive gear

On all series equipped with D361, DT361, D407 and DT407 engines, remove crankshaft pulley retaining nut, attach a suitable puller and remove crankshaft pulley. NOTE: Attach puller to tapped holes in pulley. Remove cap screws retaining oil pan to timing gear cover and loosen the remaining oil pan cap screws. Unbolt timing gear cover from engine and side rails. Pull cover forward off the dowels and remove from engine. Use care not to damage oil pan gasket. Reinstall timing gear cover by reversing the removal procedure. Tighten crankshaft pulley retaining nut to a torque of 205 ft.-lbs.

TIMING GEARS

Series 706-2706 (D282 Engine)

110. **CRANKSHAFT GEAR.** Crankshaft gear is keyed and press fitted to the crankshaft. The gear can be removed using a suitable puller after first removing the timing gear cover as outlined in paragraph 109.

Before installing, heat gear in oil, then drift heated gear on crankshaft. Make certain timing marks are aligned as shown in Fig. IH172.

111. **CAMSHAFT GEAR.** Camshaft gear is keyed and press fitted on the camshaft. Backlash between camshaft gear and crankshaft gear should be 0.0032-0.0076. Camshaft gear can be removed using a suitable puller after first removing the timing gear cover as outlined in paragraph 109 and the gear retaining nut.

Before installing, heat gear in oil until gear will slide on shaft. Install lock and nut, then tighten nut to a torque of 115 ft.-lbs. Make certain timing marks are aligned as shown in Fig. IH172.

112. **IDLER GEAR.** To remove the idler gear, first remove the timing gear cover as outlined in paragraph 109. Idler gear shaft is attached to front of engine by a cap screw.

Idler gear shaft diameter should be 2.0610-2.0615 and clearance between shaft and renewable bushing in gear should be 0.0015-0.0045. End clearance of gear on shaft should be 0.009-0.013. Make certain that oil passage in shaft is open and clean.

When reinstalling, be sure that dowel in shaft engages hole in engine front plate. The shaft retaining cap screw should be torqued to 85-90 ft.-lbs. Make certain that timing marks are aligned as shown in Fig. IH172.

113. **INJECTION PUMP DRIVE GEAR.** To remove the injection pump drive gear, first remove the timing

gear cover as outlined in paragraph 109. Remove the thrust plunger and spring from injection pump drive shaft. On early production D282 engines, remove the three cap screws and pull pump drive gear from hub. On later production D282 engines, remove nut and washer, attach a suitable puller and remove drive gear from pump shaft.

NOTE: Use caution not to pull pump shaft from injection pump. If shaft is pulled from pump, remove pump from tractor, renew shaft seals and install shaft in pump so that dimple in tang end of shaft mates with similar dimple in injection pump rotor.

When reassembling, make certain all timing marks are aligned as shown in Fig. IH172. Refer to paragraph 172 or 174 for injection pump timing.

Series 706-2706-756-2756 (D310 Engine)

114. **CRANKSHAFT GEAR.** Crankshaft gear is a shrink fit on crankshaft. To renew the gear, it is recommended that the crankshaft be removed from engine. Then, using a chisel and hammer, split the gear at its timing slot.

The roll pin for indexing crankshaft gear on crankshaft must protrude approximately 5/64-inch. Heat new gear to 400° F. and install it against bearing journal.

When reassembling, make certain all timing marks are aligned as shown in Fig. IH173.

115. **CAMSHAFT GEAR.** Camshaft gear is a shrink fit on camshaft. To renew the gear, remove camshaft as outlined in paragraph 124. Gear can now be pressed off in conventional manner, using care not to damage the tachometer drive slot in end of camshaft. When reassembling, install thrust plate and Woodruff key. Heat gear to 400° F. and install it on camshaft. NOTE: When sliding gear on camshaft, set thrust plate clearance at 0.004-0.017. Install camshaft assembly and make certain all timing marks are aligned as shown in Fig. IH173.

116. **IDLER GEAR.** To remove the idler gear, first remove timing gear cover as outlined in paragraph 109. Idler gear shaft is attached to front of engine by a special (left hand thread) cap screw. Idler gear is equipped with two renewable needle bearings. A spacer is used between the bearings.

When installing idler gear, align all timing marks as shown in Fig. IH173. Coat threads of special cap screw with "Grade B Loctite" and tighten cap screw to a torque of 67 ft.-lbs. End

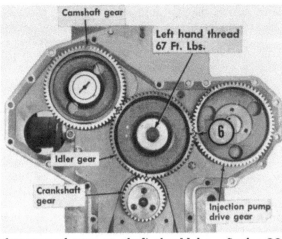

Camshaft gear

Left hand thread
67 Ft. Lbs.

Idler gear

Crankshaft
gear

6

Injection pump
drive gear

Fig. IH173 — Gear train and timing marks on series 706, 2706, 756 and 2756 equipped with D310 engines. Use timing dot next to number 6 on injection pump drive gear. Some early production gears were marked 706.

clearance of gear on shaft should be 0.008-0.013.

Timing gear backlash should be as follows:

Idler to crankshaft gear,
  New gears 0.007-0.015
  Used gears (max.) 0.0295
Idler to camshaft gear,
  New gears 0.0035-0.0107
  Used gears (max.) 0.0215
Idler to injection pump gear,
  New gears 0.0021-0.012
  Used gears (max.) 0.024

117. INJECTION PUMP DRIVE GEAR. To remove the pump drive gear, first remove timing gear cover as outlined in paragraph 109. Remove pump drive shaft nut and washer and the three hub cap screws. Attach puller (IH tool No. FES 111-2 or equivalent) to threaded holes in gear and pull gear and hub from shaft.

When reassembling, make certain all timing marks are aligned as shown in Fig. IH173. Use timing dot next to number 6 on injection pump drive gear. NOTE: Some early production gears used on series 706 and 2706 were marked 706. Refer to paragraph 173 and retime injection pump. Tighten pump drive shaft nut to a torque of 47 ft.-lbs. and the three hub cap screws to a torque of 17 ft.-lbs.

Series 806-2806-856-2856-1206-21206-1256-21256-1456-21456 Diesel

118. CRANKSHAFT GEAR. The crankshaft gear is keyed and press fitted to the crankshaft. The gear can be removed using a suitable puller after first removing the timing gear cover as outlined in paragraph 109.

Before installing, heat gear in oil, then drift heated gear on crankshaft. Make certain timing marks are aligned as shown in Fig. IH174.

119. CAMSHAFT GEAR. The camshaft gear is keyed and press fitted to camshaft. The camshaft gear can be removed using a suitable puller after timing gear cover is removed as outlined in paragraph 109, and the retaining nut not removed.

Backlash between the camshaft gear and crankshaft gear should be 0.003-0.012 with a maximum allowable backlash of 0.016.

Before installing, heat gear in oil until gear will slide on shaft. Install and tighten gear retaining nut to a torque of 55 ft.-lbs. Make certain timing marks are aligned as shown in Fig. IH174.

120. IDLER GEAR. To remove the idler gear, first remove timing gear cover as outlined in paragraph 109.

The idler gear rotates on two taper roller bearings and idler gear shaft is attached to front of cylinder block by a **left hand thread**, twelve point head cap screw. Removal of idler gear, bearings, bearing spacer and shaft is obvious.

The idler gear and idler gear shaft are available separately but the two taper bearings, two bearing cups and the bearing spacer must be renewed as an assembly.

When reinstalling, make certain timing marks are aligned as shown in Fig. IH174 and tighten the idler gear shaft cap screw to a torque of 88 ft.-lbs.

121. INJECTION PUMP DRIVE GEAR. To remove the injection pump drive gear, first remove the timing gear cover as outlined in paragraph 109.

On models equipped with the IH model RD injection pump, the drive gear can be removed from injection pump drive hub by removing the three retaining cap screws.
NOTE: Before removing pump drive gear, either position engine in the number 1 cylinder firing position outlined in paragraph 172, or mark the gear and hub so gear can be reinstalled on hub in its original position.

On models equipped with Roosa-Master injection pump, position engine in number 1 cylinder firing position as outlined in paragraph 174. Then remove nut and washer, attach a suitable puller and remove the gear.

Reassemble by reversing the disassembly procedure and align timing marks as shown in Fig. IH174.

CAMSHAFT AND BEARINGS
Series 706-2706 (D282 Engine)

122. CAMSHAFT. To remove the camshaft, first remove timing gear cover as outlined in paragraph 109. Remove rocker arm cover, rocker arms assembly and push rods. Drain and remove oil pan and oil pump. Remove engine side cover and remove cam followers (tappets). Working through openings in camshaft gear, remove the camshaft thrust plate retaining cap screws. Carefully withdraw camshaft from engine.

Recommended camshaft end play of 0.002-0.010 is controlled by the thrust plate.

Check the camshaft against the values which follow:
Journal Diameter
  No. 1 (front) 2.109-2.110
  No. 2 ....2.089-2.090
  No. 3 ....2.069-2.070
  No. 4 ....1.499-1.500

CA

IN

ID

CR

Fig. IH174 — Gear train and timing marks on D361, DT361, D407 and DT407 engines. Injection pump drive gear is different on some early D361 engines.

CA. Camshaft gear
CR. Crankshaft gear
ID. Idler gear
IN. Injection pump drive gear

When installing the camshaft, reverse the removal procedure and make certain timing marks are aligned as shown in Fig. IH172. Tighten camshaft thrust plate retaining cap screws to a torque of 38 ft.-lbs.

123. **CAMSHAFT BEARINGS.** To remove the camshaft bearings, first remove the engine as outlined in paragraph 86 and camshaft as in paragraph 122. Unbolt and remove clutch, flywheel and engine rear support plate. Remove expansion plug from behind camshaft rear bearing and remove the bearings.

Using a closely piloted arbor, install new bearings so that oil holes in bearings are in register with oil holes in crankcase. The chamfered end of bearings should be installed towards rear of engine.

Camshaft bearings are pre-sized and if carefully installed should need no final sizing. Camshaft bearing journals should have a diametral clearance in the bearings of 0.0005-0.005.

When installing plug at rear camshaft bearing, use sealing compound on plug and bore.

Series 706-2706-756-2756 (D310 Engine)

124. **CAMSHAFT.** To remove the camshaft, first remove timing gear cover as outlined in paragraph 109. Remove rocker arm cover, rocker arms assembly and push rods. Remove engine side cover and secure cam followers (tappets) in raised position with clothes pins or rubber bands. Working through openings in camshaft gear, remove camshaft thrust plate retaining cap screws. Carefully withdraw camshaft assembly.

Recommended camshaft end play is 0.004-0.017. Camshaft bearing journal diameter should be 2.2823-2.2835 for all journals.

Install camshaft by reversing the removal procedure. Make certain timing marks are aligned as shown in Fig. IH173.

125. **CAMSHAFT BEARINGS.** To remove the camshaft bearings, first remove the engine as outlined in paragraph 86 and camshaft as in paragraph 124. Unbolt and remove clutch, flywheel and engine rear support plate. Remove expansion plug from behind camshaft rear bearing and remove the bearings.

NOTE: Camshaft bearings are furnished semi-finished and must be align reamed after installation to an inside diameter of 2.2844-2.2856.

Install new bearings so that oil holes in bearings are in register with oil holes in crankcase.

Normal operating clearance of camshaft journals in bearings is 0.0009-0.0033. Maximum allowable clearance is 0.006.

When installing expansion plug at rear camshaft bearing, apply a light coat of sealing compound to edge of plug and bore.

Series 806-2806-856-2856-1206-21206-1256-21256-1456-21456 Diesel

126. **CAMSHAFT.** To remove the camshaft, first remove timing gear cover as outlined in paragraph 109. Remove rocker arm cover, rocker arms assembly and push rods. Unbolt and remove the externally mounted oil pump and the engine side covers. Raise and secure cam followers (tappets) with clothes pins or magnets. Working through openings in camshaft gear, remove camshaft thrust plate retaining cap screws. Carefully withdraw camshaft assembly.

Recommended camshaft end play of 0.002-0.010 is controlled by the thrust plate.

Check the camshaft against the values which follow:
Journal Diameter
No. 1 (front) 2.429-2.430
No. 2 ....2.089-2.090
No. 3 ....2.069-2.070
No. 4 ....1.499-1.500
When installing camshaft, reverse the removal procedure and make certain timing marks are aligned as shown in Fig. IH174. Tighten camshaft thrust plate retaining cap screws to a torque of 33 ft.-lbs.

127. **CAMSHAFT BEARINGS.** To remove the camshaft bearings, first remove the engine as outlined in paragraph 86 and camshaft as in paragraph 126. Unbolt and remove clutch, flywheel and engine rear support plate. Bearings can now be removed.

Using a closely piloted arbor, install new camshaft bearings so that oil holes in bearings are in register with oil holes in crankcase.

Camshaft bearings are pre-sized and should need no final sizing if carefully installed. Camshaft journals should have a diametral clearance of 0.001-0.0055 in the camshaft bearings. Maximum allowable clearance is 0.008.

PISTON AND ROD UNITS

All Diesel Models

128. Connecting rod and piston assemblies can be removed from above after removing cylinder head as outlined in paragraph 87 and oil pan as in paragraph 150.

Cylinder numbers are stamped on the connecting rod and cap. Numbers on rod and cap should be in register and face towards the camshaft side of engine. The arrow or FRONT marking stamped on the tops of pistons should be towards front of engine.

On series 706 and 2706 equipped with D282 engines, two types of connecting rod bolts may be used. The PLACE bolt has a head that is either notched or concave and the shank and thread diameter are nearly the same. This type attains its tension by bending the bolt head and should be torqued to 50 ft.-lbs. The PITCH bolt has a standard bolt head with a washer face. The thread diameter is larger than the shank. This type attains its tension by stretching of the shank and should be torqued to 45 ft.-lbs.

On series 706, 2706, 756 and 2756 equipped with D310 engines, tighten connecting rod nuts to a torque of 63 ft.-lbs.

On all series equipped with D361, DT361, D407 and DT407 engines, tighten connecting rod bolts to a torque of 105 ft.-lbs.

PISTONS, SLEEVES AND RINGS

Series 706-2706 (D282 Engine)

129. The cam ground aluminum pistons are fitted with two compression rings and one oil control ring. New pistons have a diametral clearance in new sleeves of 0.005-0.0076 when measured between piston skirt and installed sleeve at 90 degrees to piston pin.

The dry type cylinder sleeves should be renewed when out-of-round or taper exceeds 0.008. Inside diameter of new sleeve is 3.6873-3.690. With piston and connecting rod assemblies removed, use a suitable hydraulic puller to remove the cylinder sleeves.

NOTE: Early production D282 engines with serial numbers 71777 and below are equipped with light press fit sleeves. A pressure of 25 to 75 pounds is required to press these sleeves into position in cylinder block. Late production D282 engines with serial numbers 71778 and above are equipped with heavy press fit sleeves. These sleeves require a pressure of 750 to 2500 pounds to press them into position. The heavy press fit sleeves must not be installed in early production cylinder blocks.

On early production engines, thoroughly clean all carbon and rust from cylinder bore and counterbore. Cylinder sleeves must be clean and dry. Install sleeve in cylinder and when pressed into position, measure sleeve flange height above face of cylinder block. Refer to Fig. IH175 for view showing sleeve flange height gage (IH tool No. FES 49) being used to measure sleeve flange stand-out. Sleeve flange stand-out should be 0.001-0.004. If necessary, install shims under sleeve flange to obtain correct sleeve height. Shims are available in 0.003 and 0.005 thicknesses. All sleeves should be within 0.002 of having the same stand-out. If more than one shim is used on any one sleeve, be sure the splits of shims are staggered. If sleeves are a loose fit in cylinder block (sleeve drops all the way into place), sleeves having an oversize outside diameter must be fitted in the cylinder block. Sleeves having an oversize outside diameter of 0.002 and 0.010 are available.

On late production D282 engines, five sizes of sleeves having different outside diameters are available. Numbers 1, 2 or 3 are stamped on right side of cylinder block. These numbers (one for each cylinder) indicate the class of cylinder bores. The sleeves are marked S1, S2, S3, 0.002 oversize and 0.010 oversize. The "S1" sleeve fits a Class 1 cylinder bore, "S2" sleeve fits a Class 2 bore and "S3" sleeve fits the Class 3 bore. When installing the 0.002 oversize or 0.010 oversize sleeve, the cylinder block bore will have to be bored and/or honed to the correct oversize.

Sleeve outside diameters and cylinder block bores are as follows:

Sleeve Type	Sleeve O. D.
S1	3.8135-3.8139
S2	3.8140-3.8144
S3	3.8145-3.8149
0.002 Oversize	3.8165-3.8169
0.010 Oversize	3.8245-3.8249

Cylinder Bore Class	Bore I. D.
Class 1	3.8123-3.8127
Class 2	3.8128-3.8132
Class 3	3.8133-3.8137
0.002 Oversize	3.8143-3.8157
0.010 Oversize	3.8223-3.8237

Clean the sleeve and cylinder block bore with solvent and dry with compressed air. Lubricate sleeve and bore with clean diesel fuel. Using a hydraulic sleeve installing tool, press sleeve into cylinder block until sleeve flange projects 0.035-0.045 above top surface of cylinder block. Proper sleeve stand-out will result when sleeve installing and flange height gage (IH tool No. FES 24-6) is used. When the height gage bottoms against top of block, sleeve is properly installed. The force required to press this sleeve into position will vary from a minimum of 750 pounds to a maximum of 2500 pounds. To translate pounds of force into gage pressure for use with hydraulic ram, the effective area of the ram cylinders must be calculated by the formula $d^2 \times 0.784$. With the ram cylinder effective area calculated, the result can be divided into the stated pounds of force and pump gage pressure psi range determined. Using an OTC twin piston ram with each piston diameter of $1\frac{1}{2}$ inches, the following example is given: $(d^2 + d^2) \times 0.784$ equals $(2.25 + 2.25) \times 0.784$ equals 3.52 square inches effective area of both ram pistons. When 3.52 is divided into 750 (lbs. of force) a gage pressure of 213 psi is obtained. When 3.52 is divided into 2500 (lbs. of force), a gage pressure of 710 psi is obtained. Therefore, sleeve installation must occur between 213 and 710 psi pump gage pressure.

The top compression ring is a chrome ½-keystone ring and is installed taper side up.

The second compression ring is a taper face ring and is installed with largest outside diameter towards bottom of piston. Upper side of piston ring is marked TOP.

The oil control ring can be installed either side up. A coil spring expander is used with this ring.

Additional piston ring information is as follows:

Ring End Gap
Top compression 0.015-0.025
Second compression 0.010-0.020
Oil control 0.010-0.023
Ring Side Clearance
All rings 0.0025-0.0040

Series 706-2706-756-2756 (D310 Engine)

130. Pistons are not available as individual service parts, but only as matched units with the wet type sleeves. New pistons have a diametral clearance in new sleeves of 0.0039-0.0047 when measured between piston skirt and sleeve at 90 degrees to piston pin.

The wet type cylinder sleeves should be renewed when out-of-round or taper exceeds 0.006. Inside diameter of new sleeve is 3.8750-3.8754. Cylinder sleeves can usually be removed by bumping them from the bottom, with a block of wood.

Before installing new sleeves, thoroughly clean counterbore at top of block and seal ring groove at bottom. All sleeves should enter crankcase bores full depth and should be free to rotate by hand when tried in bores without seal rings. After making a trial installation without seal rings, remove sleeves and install new seal rings, dry, in grooves in crankcase. Wet lower end of sleeve with a soap solution or equivalent and install sleeve.

NOTE: The cut-outs in bottom of sleeve are for connecting rod clearance and must be installed toward each side of engine. Chisel marks are provided on top edge of cylinder sleeves to aid in correct installation. Align chisel marks from front to rear of engine.

If seal ring is in place and not pinched, very little hand pressure is required to press the sleeve completely in place. Sleeve flange should extend 0.003-0.005 above top surface of cylinder block. If sleeve stand-out is excessive, check for foreign material under the sleeve flange. The cylinder head gasket forms the upper cylinder sleeve seal and excessive sleeve stand-out could result in coolant leakage.

Pistons are fitted with three compression rings and one oil control ring. Prior to installing new rings, check top ring groove of piston by using Perfect Circle Piston Ring Gage No. 1 as shown in Fig. IH176. If one, or both, shoulders of gage touch ring land, renew the piston.

The top compression ring is a full keystone ring and it is not possible to measure ring side clearance in the groove with a feeler gage. Check fit of top ring as follows: Place ring in

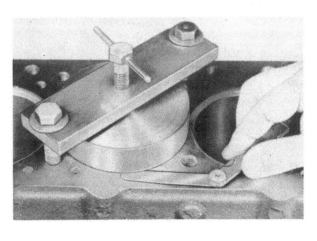

Fig. IH175 — Use sleeve flange height gage (IH tool No. FES 49) to check sleeve flange stand-out on early production D282 engines. Refer to text.

its groove and push ring into groove as far as possible. Measure the distance ring is below ring land. This distance should be 0.002-0.015. Refer to Fig. IH177 for view showing ring fit being checked using IH tools FES 68-3 and dial indicator FES 67.

The second compression ring is a taper face ring and is installed with largest outside diameter towards bottom of piston. Upper side of ring is marked TOP.

The oil control ring can be installed either side up, but make certain the coil spring expander is completely in its groove.

Additional piston ring information is as follows:

Ring End Gap
  Compression rings 0.014-0.022
  Oil control ring 0.010-0.016
Ring Side Clearance
  Top compression
    (ring drop) 0.002-0.015
  Second compression . .0.0030-0.0042
  Third compression 0.0024-0.0034
  Oil control 0.0014-0.0024

Series 806-2806-856-2856-1206-21206-1256-21256-1456-21456 Diesel

131. The cam ground pistons are fitted with two compression rings and one oil control ring. Pistons and rings are available for service in standard size only. Prior to installing new rings, check top ring groove of piston by using a Perfect Circle Piston Ring Gage No. 1 as shown in Fig. IH176. If one or both, shoulders of gage touch ring land, renew the piston.

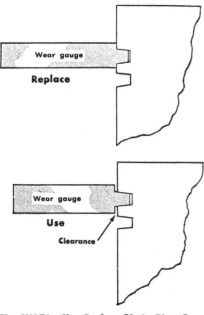

Fig. IH176—Use Perfect Circle Ring Gage No. 1 to check top ring groove of pistons used in D310, D361, DT361, D407 and DT407 engines.

The top compression ring is a full keystone ring and it is not possible to measure ring side clearance in the groove with a feeler gage. Check fit of top ring as follows: Install ring in its groove and push ring into groove as far as possible. Measure the distance ring is above or below ring land. This distance must be 0.002-0.015 below ring land on D361 and DT361 engines and 0.019 below to 0.013 above ring land on D407 and DT407 engines. See Fig. IH177 for view showing ring fit being checked using IH tools FES 68-3 and dial indicator FES 67.

The second compression ring is a taper face ring and is installed with the largest outside diameter toward bottom of piston. Upper side of ring is marked TOP.

The oil control ring is a one-piece slotted ring and can be installed either side up.

Additional piston ring information is as follows:

Ring End Gap
  Top compression ring
    D361 & DT361 . . . . . . .0.017-0.027
    D407 & DT407 . . . . .0.013-0.023
  Second compression ring
    D361 & DT361 . . . . . . .0.017-0.032
    D407 & DT407 . . . . .0.013-0.023
  Oil control ring
    D361 & DT361 . . . . . . .0.014-0.034
    D407 & DT407 . . . . .0.013-0.028
Ring Side Clearance
  Top compression ring (ring drop)
    D361 & DT361 . . −0.002 to −0.015
    D407 & DT407 . . −0.019 to +0.013
  Second compression ring
    D361 & DT361 . . . . .0.0030-0.0045
    D407 & DT407 . . . . . .0.0030-0.0048
  Oil control ring
    D361 & DT361 . . . .0.0020-0.0035
    D407 & DT407 . . . . .0.0020-0.0035

Cylinder sleeves are removed from top of cylinder block after rod and piston assemblies are out. These dry type sleeves are a tight fit in the cylinder block and must be removed and installed with a hydraulic ram. Cylinder sleeves should be renewed when out-of-round or taper exceeds 0.006. Inside diameter of new sleeves is 4.1248-4.1256 for D361 and DT361 engines and 4.3209-4.3219 for D407 and DT407 engines. New pistons should have a diametral clearance in new sleeves of 0.0038-0.0056 on D361 and DT361 engines and 0.0049-0.0069 on D407 and DT407 engines when measured between piston skirt and installed sleeve at 90 degrees to piston pin.

The three sizes of sleeves available for service are as follows: Standard outside diameter, 0.002 oversize outside diameter and 0.010 oversize outside diameter.

The fit of sleeves in cylinder block should be so that installation can be accomplished between a minimum of 300 pounds of force and a maximum of 1400 pounds of force on D361 and DT361 engines or between a minimum of 750 pounds of force and a maximum of 2000 pounds of force on D407 and DT407 engines.

To translate pounds of force into pump gage pressure for use with hydraulic ram, the effective area of the ram cylinders must be calculated by the formula $d^2 \times 0.784$. With the ram cylinder effective area calculated, the result can be divided into the stated pounds of force and pump gage pressure determined. Using an OTC twin piston ram with each piston diameter of 1½ inches, the following example is given: $(d^2 + d^2) \times 0.784$ equals $(2.25 + 2.25) \times 0.784$ equals 3.52 sq. in. effective area of both ram pistons. When 3.52 is divided into the stated pounds of force, the following pump gage pressures are obtained:

Pounds of Force	Pump Gage Pressure
300	85 psi
750	213 psi
1400	397 psi
2000	568 psi

Therefore, sleeve installation must occur between 85 and 397 psi pump gage pressure on D361 and DT361 engines and between 213 and 568 psi pump gage pressure on D407 and DT407 engines.

Before installing new sleeves, clean sleeve and cylinder block bore with

Fig. IH177—Top compression ring should be 0.002-0.015 below ring land for proper fit on D310, D361 and DT361 engines and 0.019 below to 0.013 above ring land on D407 and DT407 engines.

solvent and dry with compressed air. Lubricate sleeve and bore with clean diesel fuel. Press sleeve into cylinder block until sleeve flange projects 0.030-0.040 above top face of block on D361 and DT361 engines or 0.040-0.045 above top face of block on D407 and DT407 engines.

When installing the 0.002 oversize or 0.010 oversize sleeve, the cylinder block bore will have to be bored and/or honed to the correct oversize. Diameter of standard cylinder block bore is 4.3105-4.3119 on D361 and DT361 engines and 4.4680-4.4694 on D407 and DT407 engines.

PISTON PINS

All Diesel Models

132. The full floating type piston pins are retained in the piston bosses by snap rings. Piston pins are available in 0.005 oversize for all models except series 706, 2706, 756 and 2756 equipped with D310 engines. Specifications are as follows:

Piston pin diameter,
D282 ....1.1247-1.1249
D310 ....1.4172-1.4173
D361, DT361 & D407 ..1.4998-1.5000
DT407 (1256 & 21256) .1.4998-1.5000
DT407 (1456 & 21456) .1.6248-1.6250
Piston pin diametral clearance in piston,
D282 ....0.0000-0.0004L
D310 ....0.0001T-0.0000
D361 & DT361 ....0.0001T-0.0003L
D407 ....0.0001L-0.0005L
DT407 (1256 & 21256) ...0.0001L-0.0005L
DT407 (1456 & 21456) ...0.0005L-0.0009L
Piston pin diametral clearance in rod bushing,
D282 ....0.0005L-0.0009L
D310 ....0.0005L-0.0010L
D361 & DT361 ....0.0006L-0.0010L
D407 & DT407 ....0.0006L-0.0010L
Piston pin bushings are furnished semi-finished and must be reamed or honed for correct pin fit after they are pressed into connecting rods.

CONNECTING RODS AND BEARINGS

All Diesel Models

133. Connecting rod bearings are of the slip-in, precision type, renewable from below after removing oil pan and connecting rod caps. When installing new bearing inserts, make certain the projections on same engage slots in connecting rod and cap and that cylinder identifying numbers on rod and cap are in register and face towards camshaft side of engine. Con-

necting rod bearings are available in standard size and undersizes of 0.010, 0.020 and 0.030 for D310 engines and in standard size and undersizes of 0.002, 0.010, 0.020 and 0.030 for all other engines. Check the crankshaft crankpins and connecting rod bearings against the values which follow:

Crankpin diameter,
D282 ....2.3730-2.3740
D310 ....2.5185-2.5193
D361, DT361,
D407, DT407 ....2.9980-2.9990
Rod bearing diametral clearance,
D282 ....0.0009-0.0034
D310 ....0.0023-0.0048
D361, DT361,
D407, DT407 ....0.0018-0.0051
Rod side clearance
D282 ....0.007-0.013
D310 ....0.006-0.010
D361, DT361,
D407, DT407 ....0.004-0.015
Rod bolt (or nut) torque,
D282 PLACE bolt* ......50 ft.-lbs.
D282 PITCH bolt* ......45 ft.-lbs.
D310 ....63 ft.-lbs.
D361, DT361,
D407, DT407 ....105 ft.-lbs.
*Refer to paragraph 128 for bolt identification.

CRANKSHAFT AND MAIN BEARINGS

Series 706-2706 (D282 Engine)

134. Crankshaft is supported in four main bearings and end thrust is taken by the third (rear intermediate) bearing. Main bearings are of the non-adjustable, slip-in, precision type, renewable from below after removing the oil pan and main bearing caps. Removal of crankshaft requires R&R of engine.

Check crankshaft and main bearings against the values which follow:

Crankpin diameter 2.373-2.374
Main journal diameter 2.748-2.749
Crankshaft end play 0.005-0.013
Main bearing
diametral clearance ..0.0012-0.0042
Main bearing bolt torque ..80 ft.-lbs.

Main bearings are available in standard size and undersizes of 0.002, 0.010, 0.020 and 0.030. Alignment dowels (IH tool No. FES 6-1 or equivalent) should be used when installing the rear main bearing cap.

Series 706-2706-756-2756 (D310 Engine)

135. Crankshaft is supported in seven main bearings. Main bearings are of the non-adjustable, slip-in, precision type, renewable from below after removing the oil pan and main bearing caps. Crankshaft end

play is controlled by the flanged rear main bearing inserts. Removal of crankshaft requires R&R of engine. The crankshaft is counter-balanced by twelve weights bolted opposite to the crankshaft throws. The weights are numbered consecutively from 1 to 12 which correspond to the numbers stamped on crankshaft. The mounting holes in the balance weights are offset. The wide edge goes toward the connecting rod bearing. Balance weights are not serviced separately.

Check crankshaft and main bearings against the values which follow:

Crankpin diameter 2.5185-2.5193
Main journal diameter ..3.1484-3.1492
Crankshaft end play 0.006-0.009
Main bearing
diametral clearance ..0.0029-0.0055
Main bearing bolt torque,
Marked 10K or 10-9 .....80 ft.-lbs.
Marked 12K or 12-9 ....97 ft.-lbs.
Marked 12.9 ....140 ft.-lbs.
Balance weight bolt torque 57 ft.-lbs.

Main bearings are available in standard size and undersizes of 0.010, 0.020 and 0.030. Main bearing caps should be installed with numbered side toward camshaft side of engine.

Series 806-2806-856-2856-1206-21206-1256-21256-1456-21456 Diesel

136. Crankshaft is supported in seven main bearings and crankshaft end thrust is taken by the number seven (rear) main bearing. Main bearings are of the non-adjustable, slip-in, precision type, renewable from below after removing oil pan and main bearing caps. Removal of crankshaft requires R&R of engine. Check crankshaft and main bearings against the values which follow:

Crankpin diameter 2.9980-2.9990
Main journal diameter ..3.3742-3.3755
Crankshaft end play 0.007-0.0185
Main bearing
diametral clearance ..0.0018-0.0051
Main bearing bolt torque ..115 ft.-lbs.

Main bearings are available in standard size and undersizes of 0.002, 0.010, 0.020 and 0.030. Main bearing caps are numbered for correct location and numbered side of caps should be toward camshaft side of engine.

CRANKSHAFT SEALS

All Diesel Models

137. **FRONT.** To renew the crankshaft front oil seal, first remove hood, drain cooling system and disconnect radiator hoses. On International models, disconnect forward end of power steering cylinder from center steering arm. On Farmall models, identify and

disconnect power steering lines. On all models, identify and disconnect hydraulic oil cooler lines. Plug or cap openings to prevent dirt or other foreign material from entering hydraulic system. Support tractor under clutch housing. On Farmall models equipped with wide front axle and all International models, disconnect stay rod ball. Attach hoist to front support, unbolt front support from side rails and roll complete front assembly forward away from tractor. Remove belt, or belts, from crankshaft pulley.

On series 706 and 2706 equipped with D282 engines, remove crankshaft pulley retaining nut and clamp a bearing splitter deep in the fan drive belt groove of the crankshaft pulley. Attach a puller to bearing splitter and remove crankshaft pulley. Remove and renew seal in the conventional manner. Drive seal in until it is seated against shoulder in timing gear cover. Inspect pulley seal surface and renew or recondition pulley if surface is worn or pitted. When reinstalling crankshaft pulley, tighten pulley retaining nut to a torque of 95 ft.-lbs.

On series 706, 2706, 756, and 2756 equipped with D310 engines, remove three cap screws, flat washer and pressure ring from crankshaft. Tap crankshaft pulley with a plastic hammer to loosen pulley, then slide pulley off the wedge rings. Remove wedge rings and oil seal wear ring from crankshaft. Remove and renew oil seal in conventional manner. Renew "O" ring on crankshaft. Inspect wear ring and timing pin for wear or other damage and renew as necessary. Use a non-hardening sealer on timing pin. When installing wear ring, timing pin must engage slot in crankshaft gear. Install one pressure ring on crankshaft with thick end against wear ring. Place wedge rings in pulley bore so the slots in rings are 90 degrees apart. Slide pulley on crankshaft, aligning slot in pulley with timing pin in wear ring. Install pressure ring, flat pressure washer and three cap screws. Tighten the cap screws evenly to a torque of 55 ft.-lbs.

On all series equipped with D361, DT361, D407 and DT407 engines, remove crankshaft pulley retaining nut, attach a suitable puller and remove crankshaft pulley. NOTE: Attach puller to tapped holes in pulley. Remove and renew oil seal in conventional manner. Check condition of the crankshaft pulley sealing surface and renew the wear ring if the surface is not perfectly smooth. When reinstalling crankshaft pulley, tighten pulley

retaining nut to a torque of 205 ft.-lbs.

On all models reassemble tractor by reversing disassembly procedure.

138. REAR. To renew the crankshaft rear oil seal, first detach (split) engine from clutch housing as outlined in paragraph 229, then remove clutch and flywheel.

On series 706 and 2706 equipped with D282 engines, the rear oil seal can be removed after collapsing same. Take care not to damage sealing surface of crankshaft when removing seal. Use seal installing tool and oil seal driver (IH tools FES 6-2 and FES 6-3 or equivalent) and drive seal in until it is flush with rear of crankcase.

On series 706, 2706, 756 and 2756 equipped with D310 engines, unbolt and remove seal retainer. Check the depth that old seal is installed in retainer, then remove seal. New oil seal may be installed in retainer in any of three locations; 1/16-inch above flush with retainer (new engine original position), flush with retainer or 1/16-inch below flush with retainer. Location of seal in reainer will depend on condition of the sealing surface on crankshaft. Use new gasket and install retainer and seal assembly. Tighten seal retainer cap screws evenly to a torque of 14 ft.-lbs.

On early production 806 and 2806 tractors equipped with D361 engines prior to engine serial No. 21597, a face type rear oil seal was used. A conversion package (IH part No. 609-332C91) consisting of a wear sleeve, lip type oil seal and seal retainer, is available for up-dating the early production engines. Refer to the following paragraph for installation procedure.

On series equipped with D361, DT361, D407, and DT407 engines, unbolt and remove oil seal retainer. Remove old seal and press new seal in retainer until front of seal is flush with front (engine side) of seal bore in retainer. To remove the oil seal wear sleeve from crankshaft, use a chisel to cut **part way** through the sleeve. This will expand the sleeve diameter enough to allow it to be removed. See Fig. IH178. When installing the wear sleeve or oil seal and retainer assembly, International Harvester recommends the use of wear sleeve and seal installation tool set (IH tool No. FES 68-2). Press the wear sleeve on crankshaft flange. NOTE: One side of wear sleeve is chamfered on the I.D. to aid in starting it on the crankshaft and the opposite end is chamfered on the O.D. to aid in sliding the oil seal over the sleeve. Install seal and retainer assembly with new gasket, coat cap

Fig. IH178—View showing method of expanding rear oil seal wear sleeve for removal from D361, DT361, D407 and DT-407 engines.

screws with non-hardening sealer, center seal on crankshaft and tighten cap screws to a torque of 19 ft.-lbs.

FLYWHEEL

All Diesel Models

139. To remove the flywheel, first split tractor as outlined in paragraph 229. Then unbolt and remove clutch assembly. Remove the cap screws and lift flywheel from crankshaft. When reinstalling, tighten flywheel retaining cap screws to the following torque values:

D282 ....*60 ft.-lbs.
D310 ....100 ft.-lbs.
D361, DT361,
   D407, DT407 ....95 ft.-lbs.
*Coat threads with non-hardening sealer.

To install a new flywheel ring gear, heat same to approximately 500 degrees F.

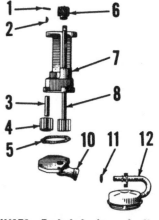

Fig. IH179—Exploded view of oil pump assembly used on D282 engines.

1.-Pin	8. Drive shaft and
2. Woodruff key	gear
3. Idler gear shaft	10. Pump cover
4. Idler gear	11. "O" ring
5. Gasket	12. Suction tube and
6. Drive gear	screen
7. Pump body	

OIL PUMP

Series 706-2706 (D282 Engine)

140. The gear type oil pump is gear driven from a pinion on the camshaft and is accessible for removal and service after removing the engine oil pan. Disassembly and overhaul of pump is obvious after an examination of the unit and reference to Fig. IH179. Gaskets (5) between pump cover and body can be varied to obtain the recommended 0.0025-0.0055 pumping (body) gear end play. Refer to the following specifications:

Pumping gears recommended
  backlash 0.003-0.006
Pump drive gear
  recommended backlash . . 0.000-0.008
Pumping gear end play 0.0025-0.0055
Gear teeth to body
  radial clearance 0.0068-0.0108
Drive shaft clearance
  in bore 0.0015-0.003
Mounting bolt torque,
  ft.-lbs. 22

Service (replacement) pump shaft and gear assembly (8) is not drilled to accept the pump driving gear pin (1). A ⅛-inch hole must be drilled through the shaft after drive gear is installed on shaft to the dimension shown in Fig. IH180.

Series 706-2706-756-2756 (D310 Engine)

141. The externally mounted gear type oil pump is located on right front side of engine and is driven by the camshaft gear. To remove the oil pump, remove two cap screws securing oil filter base to crankcase, then unbolt pump assembly. Remove pump, filter and connecting pipes as an assembly. Slide pump off the pipes.

Remove plug (16—Fig. IH181) and withdraw pressure relief valve components (12 thru 15). Remove rear cover (2); it may be necessary to tap rear cover with plastic hammer to free it from dowel pins (1). Lift out idler gear (4), then withdraw pumping drive gear (3). Remove two Wood-

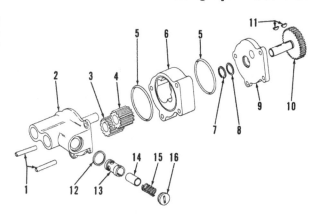

Fig. IH181 — Exploded view of oil pump assembly used on D310 engines.

1. Roll pin dowels
2. Rear cover
3. Pumping drive gear
4. Pumping idler gear
5. "O" rings
6. Pump body
7. Snap ring
8. Spacer
9. Front cover
10. Drive gear and shaft
11. Woodruff keys
12. Seal ring
13. Relief valve body
14. Relief valve piston
15. Spring
16. Plug

ruff keys (11), snap ring (7) and spacer (8) from drive gear shaft, then withdraw drive gear and shaft (10) from front cover (9). Separate front cover from pump body (6).

Inspect all parts for scoring, excessive wear or other damage. Pump covers (2 and 9) and pump body (6) are not serviced separately. Check pump against the following specifications:
Drive shaft end play
  (cover installed) 0.000-0.002
Drive shaft
  running clearance 0.001-0.0032
Idler gear to shaft
  clearance 0.001-0.0032
Pumping gears end
  clearance 0.002-0.0038
Pumping gears to body
  radial clearance 0.007-0.012
Pressure regulating spring,
  Spring free length .... 2.52 in.
  Spring test and
    length .... 18-20 lbs. at 1.858 in.

Pressure regulating valve,
  Piston diameter 0.825-0.827
  Clearance in bore 0.003-0.007
Oil pressure at 2300 RPM .... 35-50 psi

Renew all "O" rings and gaskets when reassembling and reinstalling oil pump and filter assembly.

Series 806-2806-856-2856-1206-21206-1256-21256-1456-21456 Diesel

142. The oil pump used on D361, DT361, D407 and DT407 engines is an externally mounted unit having the ports so located as to prevent drainback; thus keeping the pump filled and primed. The pump is driven via an idler gear which in turn is driven from the engine camshaft.

To remove the oil pump, first clean area around pump and engine oil cooler. Disconnect tachometer cable from tachometer drive at front end of pump, then unbolt oil pump inlet

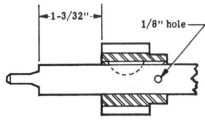

Fig. IH180—On D282 engines, new oil pump drive shaft will require a ⅛-inch hole to be drilled at location shown. Refer to text.

1-3/32"   1/8" hole

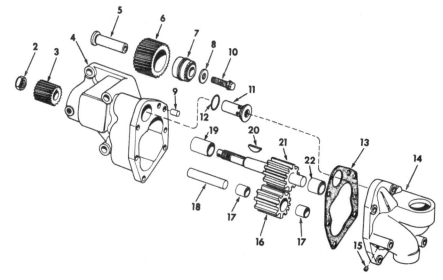

Fig. IH182—Exploded view of oil pump assembly used on D361, DT361, D407 and DT407 engines.

2. Esna nut
3. Drive gear
4. Pump body
5. Idler gear shaft
6. Idler gear
7. Bearing assembly
8. Washer
9. Dowel
10. Retaining screw
11. Safety relief valve
12. "O" ring
13. Gasket
14. Pump cover
15. Plug
16. Pumping idler gear
17. Bushings
18. Shaft
19. Bushing
20. Woodruff key
21. Pumping gear and shaft
22. Bushing

elbow from cylinder block. Unbolt and remove oil pump with pump to oil cooler tube. Remove outlet tube and inlet elbow from pump. Loosen retaining set screw and withdraw tachometer drive unit. Refer to Fig. IH182 and remove retaining screw (10), washer (8), bearing and spacer. Bump shaft (5) forward as necessary, then remove idler gear (6) and forward bearing. Remove cap screws and separate cover (14) from body (4). Withdraw safety relief valve assembly (11) and "O" ring (12). Use a wood dowel as a wedge between pumping gears (16 and 21) and remove Esna nut (2) from forward end of pump drive shaft. Remove dowel wedge and gear (16). Place pump in a press with forward end of drive shaft upward. Push drive shaft and gear (21) upward, then wedge between bottom of gear and housing as shown in Fig. IH183. Press on drive shaft until shaft has moved a distance equal to the wedge thickness (or until end of key nears bushing), then release press and add a thicker wedge. Continue this operation until key is clear of drive gear, then remove key and complete removal of shaft from gear.

NOTE: It is important that the above removal procedure be followed so that key will not enter the drive shaft bushing (19—Fig. IH182) and cause damage to bushing and/or bushing bore. The fit between drive gear and drive shaft may loosen after key is removed so use caution not to let parts fall from body.

Idler gear shaft (5) and shaft (18) can be removed from pump body, if necessary.

Refer to the following values for service information:

Pumping gears
   backlash 0.004-0.014
Pumping gears to bore
   radial clearance 0.004-0.0055
Pumping gears
   end play 0.002-0.0057
Drive shaft to
   bushing clearance 0.0025-0.004
Pumping idler gear to
   shaft clearance 0.0025-0.0035
Pump drive gear to idler
   gear backlash 0.002-0.009
Idler gear to camshaft
   backlash 0.002-0.012

Inspect all gears for wear and damaged or chipped teeth and renew as necessary. Inspect all other parts for wear, scoring or other damage. Safety relief valve (11) is available as a unit only; however, it can be disassembled, cleaned and the spring tested. Spring free length is 2.109 inches and should test 27.3 pounds when compressed to

Fig. IH183—When pressing pump shaft from drive gear, use wedges under drive gear as shown.

a length of 1.338 inches. Safety valve piston must slide freely in valve housing. Safety relief pressure is 90 psi.

NOTE: Pump safety relief valve and the oil pressure regulator valve located under the engine oil cooler rear header are similar in appearance but MUST NOT be interchanged. The oil pump safety relief valve can be identified by having an orifice in the piston and by having a heavier spring than the oil pressure regulator valve.

Reassemble pump by reversing the disassembly procedure. Tighten Esna nut (2) on drive shaft to a torque of 95 ft.-lbs. and idler gear bearing retaining screw (10) to a torque of 45 ft.-lbs. Idler gear bearing assembly (7), consisting of two bearing cups, two bearing cones and a spacer is available only as a matched assembly. Use "Plastigage" to check end play of pumping gears. Gasket (13) thickness is equal to 0.007 inch when installed. Use new "O" rings and gaskets and fill (prime) pump with new oil during installation.

OIL PRESSURE REGULATOR

Series 706-2706 (D282 Engine)

143. The D282 engines prior to serial No. 67800 are equipped with the cast iron base oil filter assembly shown in Fig. IH184. The spring loaded, plunger type oil pressure regulator valve is located in the filter base and is non-adjustable. Spring (13) should be installed with the closed coils in the plunger. Regulator valve piston (12) seals on outer diameter of valve bore. Inner end of bore is a stop for the piston and is not the valve seat. If oil pressure is lower than normal, check the spring tension and inspect valve and bore for excessive wear, scoring or other damage. Element by-pass valve is located in center tube (1). By-pass valve (3) and spring (4) can be removed from center tube after driving out pin (2). Specifications are as follows:

Pressure regulating valve,
   Valve diameter 0.743-0.745
   Clearance in bore 0.002-0.007
   Spring free length .... 3.0 in.
   Spring test and
     length .... 18 lbs. at $1\frac{13}{16}$ in.
By-pass valve,
   Spring free length .... 2 15/64 in.
   Spring test and
     length .... 3.4 lbs. at 2.0 in.
Oil pressure at 1800 RPM .. 30-40 psi

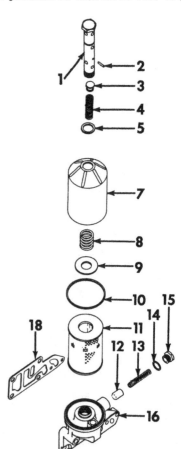

Fig. IH184 — Exploded view of cast iron base oil filter assembly used on D282 engines prior to serial No. 67800.

1. Center tube	10. Gasket
2. Pin	11. Element
3. By-pass valve	12. Regulator piston
4. Spring	13. Spring
5. Gasket	14. Gasket
7. Case	15. Plug
8. Hold-down spring	16. Filter base
9. Washer	18. Gasket

144. The D282 engines having serial No. 67800 and above are equipped with the die cast base oil filter assembly shown in Fig. IH185. The return flow check valve and element by-pass valve are located in the filter base. These valves and their springs are retained by snap rings and removal is obvious after an examination of the unit and reference to Fig. IH185. The oil pressure regulator valve assembly (6) is located in a bore in crankcase and is accessible after filter base is removed. Regulator valve is available as an asembly only; however, it can be disassembled, cleaned and inspected. See Fig. IH186.

Refer to the following specifications:

Pressure regulating valve,
    Piston diameter 0.621-0.622
    Piston clearance in bore 0.002-0.004
    Spring free length ....2.056 in.

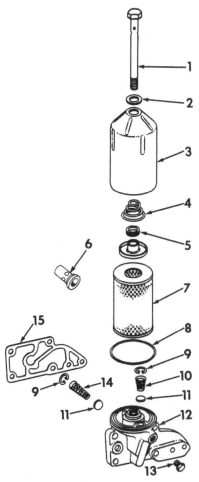

Fig. IH185—Exploded view of die cast base oil filter assembly used on D282 engines having serial No. 67800 and above. Pressure regulator valve (6) is located in crankcase behind filter base.

1. Center tube	9. Snap ring
2. Gasket	10. Check valve
3. Case	spring
4. Hold-down spring	11. Valve
5. Grommet	12. Filter base
6. Regulator valve	13. Drain plug
assembly	14. By-pass spring
7. Element	15. Gasket
8. Gasket	

Spring test and
    length ......13.85 lbs. at 1.342 in.
Check valve,
    Valve diameter 0.770
    Spring free length ......1 5/64 in.
By-pass valve,
    Valve diameter 0.770
    Spring free length ....2⅞ in.
Spring test and
    length ......4.5 lbs. at 1 5/64 in.
Oil pressure at 1800 RPM .30-40 psi

Series 706-2706-756-2756 (D310 Engine)

145. On D310 engines, the oil pressure regulator valve is located in the externally mounted oil pump. See Fig. IH181. Refer to paragraph 141 for regulator valve and spring specifications. Filter by-pass valve is located in the spin-on type oil filter. Oil pressure at 2300 RPM should be 35-50 psi.

Series 806-2806-856-2856-1206-21206-1256-21256-1456-21456 Diesel

146. The oil pressure regulator valve is located under the engine oil cooler rear header. To remove the pressure regulator, drain cooling system, remove drain plug and drain engine oil cooler. Loosen the rear header to oil cooler cap screws and pull them out as far as possible. Disconnect oil filter tube flange from rear header, then unbolt and remove rear header. Pressure regulator can now be withdrawn from its bore in crankcase.

Oil pressure regulator is available as a unit only; however, valve can be disassembled, cleaned and inspected. See Fig. IH186. Valve piston must slide freely in valve body. Check regulator valve and spring against the following specifications:

Pressure regulator valve,
    Piston diameter 0.621-0.622
    Piston clearance in bore 0.002-0.004
    Spring free length ....2.056 in.
Spring test and
    length ......13.85 lbs. at 1.342 in.
Oil pressure at 2400 RPM ..38-55 psi

NOTE: Oil pressure regulator valve and the oil pump safety relief valve located in oil pump are similar in appearance but

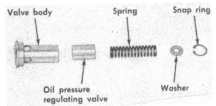

Fig. IH186—Exploded view of oil pressure regulator valve assembly used on late production D282 engines and all D361, DT361, D407 and DT407 engines.

MUST NOT be interchanged. Pressure regulator valve can be identified by having a lighter spring than the safety relief valve and by not having an orifice in the valve piston as does the safety relief valve.

OIL JET TUBES
Series 806-2806-856-2856-1206-21206-1256-21256-1456-21456 Diesel

147. Late production D361 engines and all D407 engines are equipped with six oil jet tubes located in main bearing bosses, which spray oil on the sleeves for added lubrication and cooling of pistons. Turbocharged DT361 and DT407 engines are equipped with twelve oil jet tubes.

Early production D361 engines (prior to Serial No. 6315) can be fitted with oil jet tubes; however, it is recommended that this installation be done by International Harvester as special IH tools (FES 921 drilling fixture, FES 921-1 drill bit, FES 921-2 reamer and FES 921-3 piloted driver) are required for this operation.

When overhauling engine, make certain the jet tubes are open and clean. Oil jet tubes need not be removed unless they are damaged.

OIL COOLER
Series 706-2706 (D282 Engine)

148. Oil cooler is located on left side of engine and is bolted to the oil filter base. Removal and disassembly of the unit is obvious after an examination of the unit and reference to Fig. IH188. Inspect oil cooler by-pass valve (5) for freeness in its bore. Spring (6) should have a free length of 2 27/64 inches.

Normal cleaning of the oil cooler consists of blowing out water tubes with compressed air and flushing oil passages with a suitable cleaning solvent. Renew "O" rings and gaskets when reassembling.

Series 806-2806-856-2856-1206-21206-1256-21256-1456-21456 Diesel

149. The engine oil cooler (heat exchanger) is horizontally mounted between front and rear header as shown in Fig. IH187. Service is limited to cleaning and testing although in some cases, soldering of shell seams and hubs (flanges) is permissible.

To remove the oil cooler, first drain cooling system, then remove drain plug from oil cooler and drain cooler. Remove cap screws from front header to oil cooler and the cap screw which retains front header to cylinder block. Disconnect oil pump and oil filter tube

Fig. IH187 — Left side view of D361 engine showing location of externally mounted oil pump, engine oil filters. DT361, D407 and DT407 engines are similar.

FH. Front header
OC. Oil cooler
OF. Oil filters
OP. Oil pump
RH. Rear header
TD. Tachometer drive

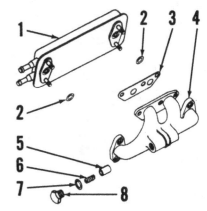

Fig. IH188—Exploded view of engine oil cooler assembly used on D282 engines.

1. Oil cooler
2. "O" rings
3. Gasket
4. Manifold
5. By-pass valve
6. Spring
7. Gasket
8. Plug

elbows. Loosen rear header to oil cooler cap screws, pull cap screws out as far as possible and remove oil cooler.

Normal cleaning of oil cooler consists of blowing out water tubes with compressed air or the use of a rotary brush or rod of proper diameter. Lubricating oil section of oil cooler can be back flushed with a suitable cleaning solvent or flushing oil. If necessary, oil cooler water tubes can be cleaned by plugging the oil ports and immersing the oil cooler in "Oakite" or similar cleaning solutions.

To test the oil cooler for internal leakage, fabricate two adapter plates with gaskets and block off oil inlet and outlet. Install an air coupling in oil drain hole and attach a regulated air supply. Immerse oil cooler in water of at least 120 degrees F. and allow oil cooler temperature to equalize with water temperature. Apply not more than 80 psi of air pressure. Any leaks present will be readily apparent. Internal leakage will require renewal of oil cooler.

Using new gaskets, reinstall oil cooler by reversing the removal procedure.

OIL PAN

All Diesel Models

150. Removal of oil pan is conventional and on tricycle model tractors can be accomplished with no other disassembly. On Farmall models equipped with adjustable wide tread front axle, the stay rod must be disconnected from stay rod support and stay rod support from side rails before removal of oil pan can be accomplished.

On International and "All Wheel Drive" models, disconnect stay rod (or bracket) from clutch housing and front axle from axle support, then raise front of tractor to provide clearance for oil pan removal.

GASOLINE FUEL SYSTEM

CARBURETOR

All Models So Equipped

151. Series 706, 2706, 756, 2756, 806, 2806, 856 and 2856 gasoline fuel tractors are equipped with International Harvester built 1⅜ inch updraft carburetors. Disassembly and overhaul is obvious after an examination of the unit and reference to Fig. IH190.

Parts data is as follows:

Series 706-2706

(C263 Engine)	IH Part No.
Carburetor	388424R94
Gasket package	387454R92
Overhaul package	387459R92
Main jet	383512R1
Idle jet	49798D70
Idle needle	362528R1
Discharge nozzle	383504R1
Venturi	47407D29
Float	47398DX
Float needle and seat	47396DAX40

Series 706-2706-756

2756 (C291 Engine)	IH Part No.
Carburetor	396197R92
Gasket package	387454R92
Overhaul package	387459R92
Main jet	376440R1
Idle jet	49798D70
Idle needle	362528R1
Discharge nozzle	383504R1
Venturi	47407D29
Float	47398DX
Float needle and seat	47396DAX40

Series 806-2806-856

2856 (C301 Engine)	IH Part No.
Carburetor	388425R94
Gasket package	387454R92
Overhaul package	387459R92
Main jet	384023R1
Idle jet	49798D70
Idle needle	362528R1
Discharge nozzle	384024R1
Venturi	47407D35
Float	47398DX
Float needle and seat	47396DAX40

152. LOW IDLE ADJUSTMENT. Before attempting to adjust the carburetor, first start engine and let it run until thoroughly warmed. Then, adjust idle speed stop screw (2—Fig. IH190) to obtain a low idle engine speed of 425 rpm. Turn idle mixture screw (10) in or out as required to obtain smoothest idle. Readjust idle speed stop screw, if necessary, to obtain correct low idle speed.

If carburetor has been disassembled, initial setting of idle mixture needle is one turn open.

153. MAIN FUEL ADJUSTMENT. To prevent engine "run-on" or "dieseling" after ignition is switched to "OFF" position, the carburetors are equipped with a solenoid fuel shut-off valve which stops fuel flow through the main jet. The main fuel adjusting screw (50—Fig. IH190) is located in outer end of solenoid unit. To adjust the main fuel screw, pro-

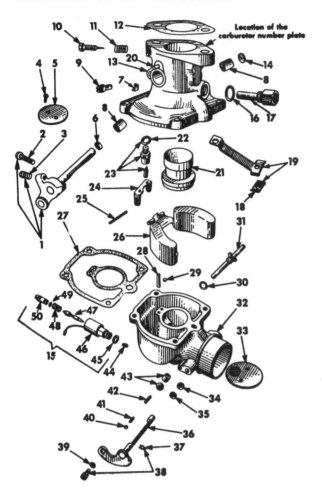

Fig. IH190 — Exploded view of carburetor typical of those used on gasoline engine tractors.

1. Throttle shaft
2. Stop screw
3. Spring
5. Throttle plate
6. Dust seal
7. Clip
8. Bushing
9. Plug
10. Idle adjusting screw
11. Spring
12. Gasket
13. Body
14. Expansion plug
17. Screen assembly
18. Clamp
19. Bracket
20. Stop pin
21. Venturi
23. Float needle and seat
24. Pivot support
25. Float pivot pin
26. Float
27. Bowl gasket
28. Idle metering jet
29. Air bleed
31. Metering nozzle
32. Bowl assembly
33. Choke plate
34. Dust seal
35. Dust seal
36. Choke shaft
37. Swivel retainer
38. Swivel
39. Washer
40. Ball
41. Friction spring
42. Stop pin
43. Drip hole filler
44. Fuel adjustment screw seat
46. Body (solenoid)
47. Needle
48. Compression spring
49. "O" ring
50. Adjusting screw

ceed as follows: With ignition switch in "OFF" position, turn adjusting screw in (clockwise) until it just contacts the solenoid plunger. Then, turn adjusting screw out 4½ turns.

154. HIGH IDLE SPEED ADJUSTMENT. The carburetor throttle lever has a stop incorporated; however, engine high idle rpm and governed rpm are controlled by the engine governor. Refer to the governor section for adjustment information.

155. FLOAT SETTING. To check the float setting, remove throttle body from bowl assembly. Invert throttle body and measure distance between top of free ends of float and gasket surface on throttle body. Distance should be 1 5/16 inches. Bend float lever, if necessary, to obtain this setting.

LP-GAS SYSTEM

All Models So Equipped

156. An Ensign model CBX 1½ inch updraft carburetor is used on all LP-Gas equipped models. Early series 706 and 2706 prior to engine serial No. 66123 and series 806 and 2806 prior to engine serial No. 5483 are equipped with Ensign model RDC regulator-vaporizer. Later series 706, 2706, 806 and 2806 and series 756, 2756, 856 and 2856 are equipped with Ensign model RDH regulator-vaporizer. The carburetor is equipped with a diaphragm economizer. Three adjustments; main (load) adjustment, starting adjustment and throttle stop (idle speed) adjustment are located on the carburetor while the idle mixture adjustment is located on the regulator. The system also incorporates a renewable cartridge fuel filter.

157. ADJUSTMENTS. Before attempting to start engine, check and be sure that initial adjustments are as follows:

Series 706-2706-756-2756
Starting adjustment ......1 turn open
Main adjustment .... 3⅝ turns open
Idle mixture
    adjustment ....1½ turns open

Series 806-2806-856-2856
Starting adjustment ......1 turn open
Main adjustment ....5⅜ turns open
Idle mixture
    adjustment ....1⅝ turns open
With initial adjustment made, start engine and run until engine reaches operating temperature. Place throttle

Fig. IH191—Exploded view of Ensign model CBX carburetor used on LP-Gas equipped tractors.

1. Economizer spring
2. Economizer diaphragm
3. Lock nut
4. Lock adjusting screw
5. Throttle plate
6. Choke plate
7. Starting adjusting screw
9. Venturi
11. Economizer bleed
12. Economizer cover
13. Tee
14. Connector

15. Balance tube
16. Elbow
17. Expansion plug
18. Gasket
19. Bushing
20. Seal
21. Throttle shaft
22. Pin
23. Dust seal
24. Choke shaft
25. Clamp
26. Bracket
27. Choke lever
28. Gasket
29. Intake elbow
30. Expansion plug
31. Valve lever

control lever in low idle position and adjust idle speed screw on carburetor to obtain an engine low idle speed of 425 rpm. Turn idle mixture needle on regulator either way as required to obtain the highest and smoothest engine operation. Readjust the carburetor idle speed screw, if necessary, to maintain engine low idle speed of 425 rpm.

With engine low idle adjusted, place throttle control in high idle position. Turn the main (load) adjustment screw (4—Fig. IH191) inward until engine begins to falter; then back-out the screw until full power is restored and engine operates smoothly.

NOTE: In some cases, it may be necessary to vary the main fuel adjustment slightly after load is placed on engine.

The initial starting adjustment should provide satisfactory starting performance; however, it may be varied if cold starting is not satisfactory.

158. CARBURETOR OVERHAUL. The carburetor is serviced similar to conventional gasoline type; that is, it can be completely disassembled, cleaned and worn parts renewed. Refer to Fig. IH191 for an exploded view of carburetor. Pay particular attention to the economizer diaphragm assembly and make certain that vacuum connections to the economizer chamber do not leak.

159. REGULATOR TROUBLE SHOOTING. To test the regulator on the engine, install a 0-30 psi test gage at the pipe plug opening (R—Fig. IH-192 or IH194). Slowly open the vapor service valve on tank. The pressure reading on test gage should raise and hold steady within the range of 8-9 psi. If the pressure is a few pounds over or under this range but remains steady, it is an indication that the high pressure valve lever requires adjustment. If test gage pressure continues to raise beyond 15 psi, the high pressure valve is leaking. High pressure valve (C—Fig. IH192 or IH194) can be serviced without removing or disassembling the regulator unit. Severe leakage at high pressure valve (pressure above 15 psi) will force low pressure valve off its seat, upsetting fuel economy, causing loss of control of idling adjustment and preventing proper starting due to an overly rich starting mixture.

Leakage through the high pressure diaphragm will have an identical effect on operation as a leaking high pressure valve; however, no increase in test gage pressure beyond the normal range (8-9 psi) may be indicated.

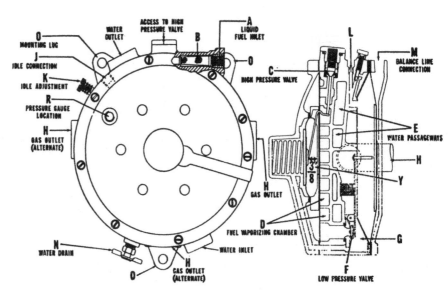

Fig. IH192—Sectional views of Ensign model RDC regulator-vaporizer used on early series 706, 2706, 806 and 2806 LP-Gas equipped tractors.

If test gage pressure holds steady within the range of 8-9 psi and there is no control of idle adjustment, improper starting is experienced and after standing awhile, frost appears on regulator, the low pressure valve (F) is leaking. A severe leak at low pressure valve can normally be heard.

Leakage through the low pressure diaphragm effects only the sensitive response of low pressure valve to the fuel demand of the engine. Response drops off rapidly with the increase in size of hole in diaphragm.

"Freeze up" of the regulator-vaporizer during engine operation is caused

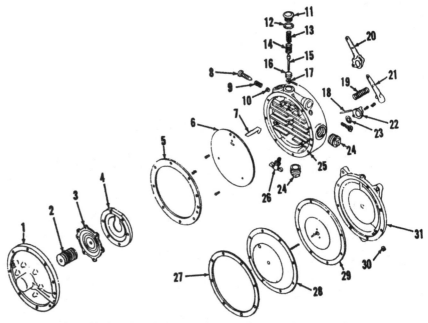

Fig. IH193—Exploded view of model RDC regulator-vaporizer.

1. Cover	12. Gasket	21. Low pressure
2. High pressure	13. High pressure	valve lever
spring	valve spring	22. Low pressure
3. High pressure	14. Seat retainer	valve seat
diaphragm	spring	23. Seal
4. Cover	15. High pressure	24. Pipe plug
5. Cover gasket	valve	25. Body
6. Partition plate	16. High pressure	26. Drain valve
7. Valve lever	valve seat	27. Gasket
8. Idle mixture	17. Seal	28. Partition plate
adjusting screw	18. Pivot pin	29. Low pressure
9. Spring	19. Low pressure	diaphragm
10. Bleed screw	valve spring	30. Pipe plug
11. Retainer	20. Low pressure	31. Support plate
	valve assembly	

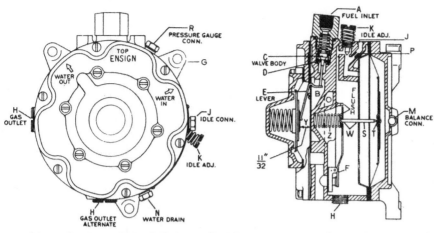

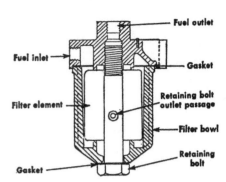

Fig. IH196—Sectional view of LP-Gas fuel filter. Some filters may have a center stud and nut instead of the retaining bolt shown.

Fig. IH194—Sectional views of Ensign model RDH regulator-vaporizer used on late series 706, 2706, 806 and 2806 and series 756, 2756, 856 and 2856 LP-Gas equipped tractors.

by lack of circulation of hot water through the regulator vaporizer unit. Check for restrictions.

160. **MODEL RDC REGULATOR OVERHAUL.** Disassembly of the regulator is obvious after an examination of the unit and reference to Figs. IH-192 and IH193. Thoroughly clean all parts and renew any which are excessively worn. When reassembling the unit, make certain that the valve levers are set to the correct dimensions as follows: The high pressure lever height dimension should be ⅜-inch from top of lever to face of partition plate as shown at (Y—Fig. IH-192). Bend lever to adjust.

A boss is machined on body (25—Fig. IH193) and is marked with an arrow for the purpose of setting the

low pressure valve. The valve lever should be centered on the arrow as the lever retaining screws are tightened and the height of the lever should be set to the height of the boss. Lever can be bent if necessary.

161. **MODEL RDH REGULATOR OVERHAUL.** Disassembly of the RDH regulator is obvious after an examination of the unit and reference to Figs. IH194 and IH195. Wash parts in clean suitable solvent and blow out all passageways with compressed air. When reassembling, renew all gaskets and any parts that show excessive wear or other damage. Adjust pressure valve levers to correct dimensions as follows: High pressure lever (E—Fig. IH194) should be set at $\frac{11}{32}$-inch from top of lever to face of plate. See dimension (Y). Bend lever, if necessary, to obtain this dimension.

When installing low pressure valve (27—Fig. IH195) and spring (25), center the low pressure lever with push pin hole in center of partition plate (30). A rib is provided in recess of body (19) for the purpose of setting low pressure lever height. Top of lever, when valve is seated, should be flush with top of this rib. Bend lever, if necessary, to obtain this setting.

162. **LP-GAS FILTER.** The filter is designed for a working pressure of 375 psi and filter element should be renewed when filter becomes clogged enough to restrict flow of fuel.

A clogged filter element causes a pressure drop within the filter and generally results in freezing (frost) at the filter as well as a noticeable drop in engine power.

Renewal of filter element is obvious after an examination of the filter and reference to Fig. IH196. Use care when handling element to prevent crushing the sides. Make certain gasket surfaces are clean and use all new gaskets when assembling.

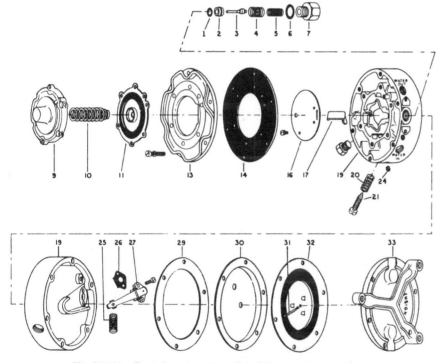

Fig. IH195—Exploded view of model RDH regulator-vaporizer.

1. "O" ring	11. High pressure	25. Valve spring
2. Valve seat	diaphragm	26. Gasket
3. High pressure	13. Cover	27. Low pressure
valve	14. Gasket	valve
4. Spring	16. Plate	29. Gasket
5. Valve spring	17. Valve lever	30. Plate
6. Gasket	19. Body	31. Push pin
7. Valve retainer	20. Spring	32. Low pressure
9. Cover	21. Idle screw	diaphragm
10. Spring	24. Bleed screw	33. Support plate

DIESEL
FUEL SYSTEM

The diesel fuel system of early series 706 and 2706 (D282 engine) and series 806 and 2806 (D361 engine) may be equipped with either a Roosa-Master fuel injection pump or an International Harvester model RD injection pump. Late series 706 and 2706 and series 756 and 2756 (D310 engine) are equipped with a Robert Bosch injection pump. Series 856 and 2856 (D407 engine), 1206 and 21206 (DT361 engine), 1256, 21256, 1456 and 21456 (DT407 engine) are all equipped with a Roosa-Master injection pump.

The injectors differ between the series 706 and 2706 equipped with the D282 engine and all other series in that the D282 engine is equipped with pre-combustion chambers and glow plugs. All other engines are of the direct injection type.

When servicing any unit of the diesel system, the maintenance of absolute cleanliness is of utmost importance. Of equal importance is the avoidance of nicks or burrs on any of the working parts.

Probably the most important precaution that service personnel can impart to owners of diesel powered tractors, is to urge them to use an approved fuel that is absolutely clean and free from foreign material. Extra precaution should be taken to make certain that no water enters the fuel storage tanks.

FILTERS AND BLEEDING

All Diesel Models

163. All diesel tractors are equipped with two fuel filters. On series 806 and 2806 tractors equipped with the IH model RD injection pump, the primary filter is the front of the two filters. On all tractors equipped with Robert Bosch and Roosa-Master injection pumps, the primary filter is the rear of the two filters. Series 856, 2856, 1256, 21256, 1456 and 21456 tractors are equipped with spin-on cartridge type fuel filters, whereas on all other series the fuel filters are fitted with renewable elements. Filters should be serviced when engine shows signs of losing power.

164. To bleed the fuel system on series 806 and 2806 tractors equipped with the IH model RD injection pump, proceed as follows: Refer to Fig. IH-197 and open vent valves on top of both filters. Push in on the priming valve, located on bottom rear of injection pump as shown in Fig. IH198, and rotate valve until it locks in the

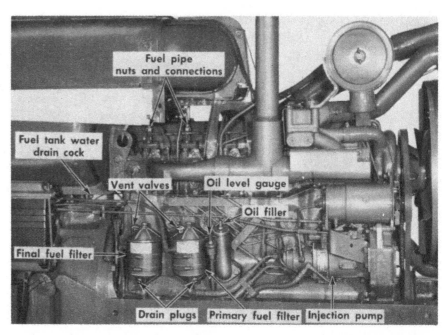

Fig. IH197—Right side view of series 806 diesel engine equipped with IH model RD injection pump, showing location of diesel fuel system components. Note location of primary and final fuel filters.

open (in) position. When bubble free fuel flows from both filters, close vent valves. Then, rotate priming valve until it snaps to closed (out) position. Fuel system is now free of air and engine should be ready to start.

165. To bleed the fuel system on series 706, 2706, 756 and 2756 D310 engine equipped with the Robert Bosch injection pump, refer to Fig. IH199 and proceed as follows: Open large vent valve at top rear of primary filter and the two small vent valves on final filter. When the out-

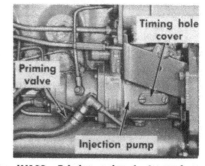

Fig. IH198—Priming valve is located at bottom rear of IH RD injection pump.

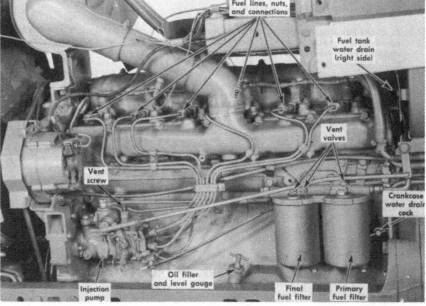

Fig. IH199—Left side view of series 706, 2706, 756 and 2756 diesel 310 engine equipped with Robert Bosch injection pump, showing location of diesel fuel system components.

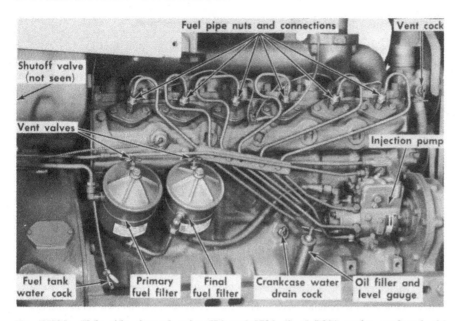

Fig. IH200—Right side view of series 706 and 2706 diesel D282 engine equipped with Roosa-Master injection pump, showing location of diesel fuel system components.

flowing fuel is free of air bubbles, close vent valves. Next, open vent screw on injection pump until fuel is flowing free of air bubbles, then retighten vent screw. Fuel system is now free of air.

166. To bleed the fuel system on all series equipped with the Roosa-Master injection pump, refer to Figs. IH200, IH201 and IH202 and proceed as follows: Loosen vent valve on top of primary filter and allow fuel to run until a solid stream with no air bubbles appears; then, close vent valve. Repeat this operation on the final fuel filter. Loosen the fuel supply line connection on the injection pump inlet elbow and when bubble free fuel appears, tighten connection. Fuel system is now free of air.

INJECTION PUMP

The International Harvester model RD, Robert Bosch and Roosa-Master injection pumps are all of the rotary distributor type. Because of the special equipment needed, and skill required of servicing personnel, service of injection pumps is generally beyond the scope of that which should be attempted in the average shop. Therefore, this section will include only timing of pump to engine, removal and installation and the linkage adjustments which control the engine speeds.

If additional service is required, the pump should be turned over to an International Harvester facility which is equipped for diesel service, or to some other authorized diesel service station. Inexperienced personnel should NEVER attempt to service diesel injection pumps.

REMOVE AND REINSTALL PUMP

Series 706-2706-806-2806 (IH Model RD)

167. Prior to removal of injection pump thoroughly clean injection pump, fuel lines and side of engine. Remove timing hole cover on right side of clutch housing as shown in Fig. IH203. Using a bar on flywheel teeth, turn engine in normal direction of rotation until number one piston is 50 degrees BTDC on series 806 and 2806

or 40 degrees BTDC on series 706 and 2706 on compression stroke. Shut off fuel and disconnect supply and return lines. On series 806 and 2806, remove hood, exhaust manifold and heat shields, generator and generator bracket. On all models, disconnect control rod from injection pump. Disconnect lines from injectors and immediately cap the openings. Unbolt and remove injection pump drive gear cover from front of timing gear cover, mark gear and hub, then remove the cap screws which retain drive gear to pump drive hub. Unbolt injection pump from engine and withdraw injection pump and injector lines. Injector lines can now be removed from pump.

Before reinstalling pump, remove timing hole cover from side of pump and make certain the "X" marked end of governor shaft is visible in opening. See Fig. IH204. Mount pump on engine, align marks of drive gear and hub and install gear retaining cap screws. Recheck pump timing as outlined in paragraph 172. Complete balance of reassembly by reversing the disassembly procedure. Bleed fuel system as outlined in paragraph 164.

Series 706-2706-756-2756 (Robert Bosch)

168. To remove the Robert Bosch injection pump from the D310 engine, first thoroughly clean injection pump, fuel lines and side of engine. Shut off fuel at tank and remove timing hole

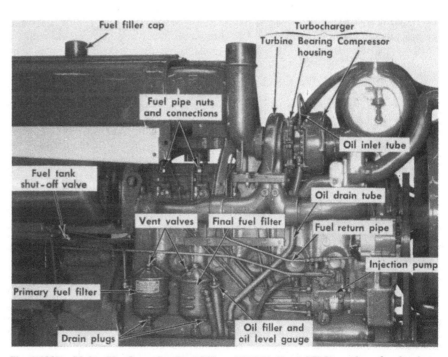

Fig. IH201—Right side view of series 1206 and 21206 diesel DT361 engine, showing location of diesel fuel system components. Location of primary and final fuel filters and injection pump are the same on series 806 and 2806 D361 engine equipped with Roosa-Master injection pump.

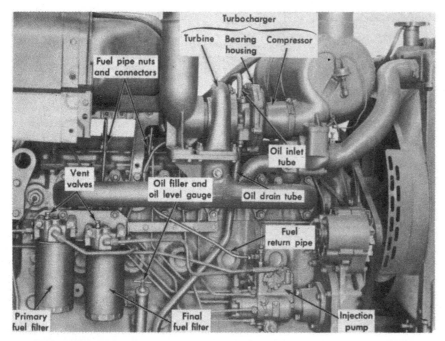

Fig. IH202—Right side view of series 1256, 21256, 1456 and 21456 diesel DT407 engine, showing location of diesel fuel system components. Fuel system components are located the same on naturally aspirated D407 engine used on series 856 and 2856. Note the spin-on type fuel filters.

Fig. IH205—View of timing pointer (TP) and timing line (TL) on Robert Bosch injection pump.

cover screws from pump. Rotate cover 90 degrees and remove by prying evenly with two screwdrivers. Remove timing hole cover on right side of clutch housing as shown in Fig. IH203. Using a bar on flywheel, turn engine in normal direction of rotation until timing line (TL—Fig. IH205) on face cam is aligned with timing pointer (TP) in pump.

NOTE: The face cam has two timing lines. Near one of the lines, a letter "L" is etched on face cam. DO NOT use this timing line.

Disconnect throttle control rod and shut-off cable from pump. Disconnect fuel inlet and return lines from pump, then remove high pressure injection lines. Immediately cap or plug all openings. Unbolt and remove the rectangular cover from timing gear cover. Remove nut and washer from pump

drive shaft, then remove three cap screws securing hub to drive gear. Remove nuts from pump mounting studs. Install a puller (IH tool No. FES 111-2 or equivalent) to tapped holes in drive gear and force drive shaft from hub. Remove injection pump.

CAUTION: Do not turn crankshaft while pump is removed.

When reinstalling pump, make certain that timing pointer (TP) and timing line (TL) are aligned, then reinstall pump by reversing removal procedure. Align scribe mark on pump mounting flange with punch mark on engine front plate. Tighten pump drive shaft nut to a torque of 47 ft.-lbs. and the three hub to drive gear cap screws to a torque of 17 ft.-lbs.

Bleed fuel system as in paragraph 165, then check and adjust injection pump timing as in paragraph 173.

Series 706-2706 (Roosa-Master)

169. To remove the Roosa-Master injection pump from the D282 engine, first clean injection pump, fuel lines and side of engine. Remove timing hole cover on right side of clutch housing as shown in Fig. IH203. Using a bar on flywheel teeth, turn engine in normal direction of rotation until the TDC mark on flywheel is aligned with timing pointer. Shut off fuel and remove timing hole cover from injection pump. Both timing marks (TM— Fig. IH206) should be visible through opening. If only one mark is visible, rotate crankshaft one complete revolution. Disconnect the control rod and the supply and return lines from injection pump. Disconnect lines from injectors and immediately cap or plug

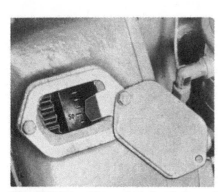

Fig. IH203 — View of timing hole and pointer located on right side of clutch housing of series 806 and 2806 diesel tractors. Other models are similar.

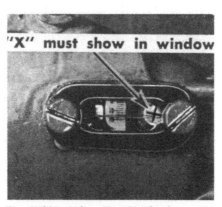

Fig. IH204—When IH RD injection pump is properly timed, the centerline of the "X" marked end of the governor shaft will be aligned with the scribe line at front of opening.

Fig. IH206—Timing marks (TM) on governor drive plate and cam ring on Roosa-Master injection pump.

all openings. Remove injection pump drive gear cover from timing gear cover and withdraw thrust plunger and spring from end of pump drive shaft.

On early production D282 engines, mark relative position of drive gear on hub, then remove the three cap screws (S—Fig. IH207) which secure gear to hub. Unbolt pump mounting adapter from engine front plate and withdraw pump with drive shaft and hub assembly.

On later production D282 engines, remove nut and washer from front end of pump drive shaft. Unbolt pump mounting adapter from engine front plate. Using a suitable puller, force pump shaft rearward from drive gear and remove pump and drive shaft assembly.

On all engines, pump drive gear will remain in the timing gear cover and cannot be removed until timing gear cover is removed. Do not rotate crankshaft while pump is removed.

NOTE: If pump drive shaft should happen to be pulled from pump, renew the two opposed seals on shaft. Carefully install shaft in pump, making certain that dimple on shaft tang is in register with dimple in slot in the pump rotor. Be sure lip of rear seal is not rolled back during installation of shaft.

Before reinstalling pump, make certain pump timing marks (TM—Fig. IH206) are aligned. Install pump by reversing removal procedure. Check and adjust pump timing as outlined in paragraph 174. Bleed fuel system as in paragraph 166.

Fig. IH207—Injection pump drive gear is secured to pump drive shaft hub by three cap screws (S) on early production D282 engines equipped with the Roosa-Master pump.

Series 806-2806-856-2856-1206-21206-1256-21256-1456-21456 (Roosa-Master)

170. To remove the Roosa-Master injection pump from the D361, DT361, D407 and DT407 engines, first clean injection pump, fuel lines and side of engine. Remove hood and on turbocharged models, carefully remove turbocharger assembly as outlined in paragraph 191. On all models, unbolt and remove exhaust manifold. Shut off fuel and remove timing hole cover from injection pump. Remove timing hole cover on right side of clutch housing as shown in Fig. IH203. Using a pry bar on flywheel teeth, turn crankshaft in normal direction of rotation until timing marks (TM—Fig. IH206) are aligned in pump timing window. Disconnect control rod and the supply and return lines from pump and the lines from injectors. Cap or plug all openings immediately. Wire the throttle lever on pump in high idle (full fuel) position. This will allow governor spring tension to hold governor weights in position. Unbolt injection pump from pump drive adapter, then withdraw pump with injector lines from pump shaft.

If necessary, the injection pump drive shaft can be removed as outlined in paragraph 171. When reinstalling injection pump, align the timing slot on pump drive shaft with the timing pin on the pump rotor. Carefully slide pump into position. Align timing marks (TM—Fig. IH206) and tighten the pump mounting nuts. The balance of reassembly is the reverse of disassembly procedure. Bleed fuel system as in paragraph 166 and check and adjust injection pump timing as in paragraph 174.

171. To remove the injection pump drive shaft assembly, first remove injection pump as in paragraph 170. Unbolt and remove pump drive gear cover from front of timing gear cover. Remove the three cap screws securing pump adapter to engine front plate. Remove nut and washer from front end of pump drive shaft and using a suitable puller, force shaft rearward from gear. Withdraw pump shaft and adapter assembly.

Any further disassembly will be obvious after an examination of the unit and reference to Fig. IH208.

CAUTION: Do not rotate crankshaft while pump shaft and adapter assembly is removed.

When reinstalling tighten gear retaining nut on drive shaft to a torque of 55 ft.-lbs. Refer to paragraph 170 for pump installation procedure.

INJECTION PUMP TIMING

Series 706-2706-806-2806 (IH Model RD)

172. **STATIC TIMING.** To check or adjust the injection pump timing, proceed as follows: Remove timing hole cover on right side of clutch housing as shown in Fig. IH203. Using a pry bar on flywheel teeth, turn crankshaft in normal direction of rotation until the 40 degree mark (706 and 2706) or 50 degree mark (806 and 2806) on flywheel is aligned with timing pointer. Shut off fuel and remove timing hole cover from side of injection pump. The "X" marked end of governor shaft should be visible through timing hole as shown in Fig. IH204. If not, rotate crankshaft one complete revolution and again align the correct degree mark with timing pointer.

The centerline of the "X" on end of governor weight shaft should be aligned with the scribe line on front side of pump timing hole opening. If timing marks are not aligned, remove injection pump drive gear cover from front of timing gear cover, loosen the three cap screws which retain drive gear to pump shaft hub and rotate hub until timing marks are aligned. Retighten gear retaining cap screws and install all removed covers.

Series 706-2706-756-2756 (Robert Bosch)

173. **STATIC TIMING.** To check or adjust static timing, first shut off fuel at tank and remove pump timing hole screws. Rotate cover 90 degrees and remove cover by prying evenly with two screwdrivers. Remove timing hole cover from right side of clutch housing as shown in Fig. IH203. Using a pry bar on flywheel teeth, turn crank-

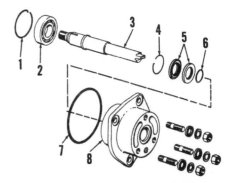

Fig. IH208—Exploded view of Roosa-Master injection pump drive shaft assembly used on D361, DT361, D407 and DT407 engines.

1. Bearing retaining ring	5. Shaft seals
2. Bearing	6. "O" ring
3. Drive shaft	7. "O" ring
4. Seal retaining ring	8. Adapter

shaft in normal direction of rotation until number one piston is coming up on compression stroke. Continue turning crankshaft until the 14 degrees BTDC mark on flywheel is aligned with timing pointer on clutch housing. At this time, timing line (TL—Fig. IH205) on face cam should be aligned with timing pointer (TP) on roller retainer ring. If not, first make certain that scribe line on pump mounting flange is aligned with punch mark on engine front plate. Then, unbolt and remove rectangular cover from front of timing gear cover. Loosen the three cap screws securing pump drive gear to pump shaft hub. Rotate hub as required to align timing line on face cam with the timing pointer. Tighten hub retaining cap screws to a torque of 17 ft.-lbs. and install all removed covers.

All Series Equipped With Roosa-Master Pump

174. STATIC TIMING. To check or adjust static timing, first shut off fuel at tank and remove timing hole cover from side of injection pump. Remove timing hole cover from right side of clutch housing as shown in Fig. IH203. Using a pry bar on flywheel teeth, turn crankshaft in normal direction of rotation until number one piston is coming up on compression stroke. Continue turning crankshaft until the following degree mark on flywheel is aligned with timing pointer on clutch housing:

706-2706 . . . . .TDC
806-2806-1206-21206 . . . . . . .8° BTDC
856-2856-1256-21256 . . . . . . .6° BTDC
1456-21456 . . . . .2° ATDC

At this time, timing marks on injection pump cam ring and governor drive plate should be aligned as shown in Fig. IH206. If timing marks are not aligned as shown, loosen the pump mounting stud nuts and rotate pump as required to align the marks. Tighten the mounting nuts and reinstall the removed covers.

INJECTION PUMP SPEED ADJUSTMENT

Series 706-2706-806-2806 (IH Model RD)

175. LOW & HIGH IDLE. To adjust the low idle speed, start engine and bring to normal operating temperature. Move throttle control lever to low idle position. Disconnect control rod from pump control lever. Move pump control lever toward shut-off position until the internal shut-off cam is felt. Adjust control rod until rod end aligns with hole in pump lever. Then shorten the control rod ⅛-inch. Connect control rod and check engine low idle speed which should be 625-675 rpm. Move throttle control lever to shut-off position. In this position, pump control lever should contact the internal stop and flex the control lever spring about 5 degrees. To check engine high idle speed, place throttle control lever in high idle position. Pump control lever should contact its stop and flex the spring approximately 5 degrees. At this position, engine high idle should be 2615-2620 rpm on series 806 and 2806 or 2507 rpm on series 706 and 2706.

Series 706-2706-756-2756 (Robert Bosch)

176. LOW & HIGH IDLE. To adjust low idle speed, start engine and bring to operating temperature. Place throttle control lever in low idle position, loosen jam nut and turn low idle speed stop screw (L—Fig. IH209) as required to obtain an engine low idle speed of 650 rpm. Tighten jam nut.

Move throttle control lever to high idle position, loosen jam nut and turn high idle stop screw (H) as required to obtain an engine high idle speed of 2530 rpm. Tighten jam nut.

With engine high idle speed properly adjusted, the rated load engine speed should be 2300 rpm. To adjust rated load rpm, use a dynamometer and load the engine to maintain rated rpm with throttle control lever in high idle position. Adjust the maximum fuel stop screw to obtain approximately 76 (pto) horsepower at rated load (2300) rpm. CAUTION: Do not overfuel the engine or attempt to increase horsepower above the rated load.

177. SHUT-OFF PLUNGER. The shut-off plunger (S—Fig. IH209) provides an excess fuel starting position for the shut-off control lever on injection pump. To adjust the plunger,

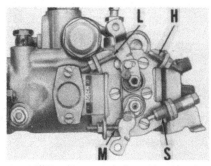

Fig. IH209—View of Robert Bosch injection pump showing adjustment points.

H. High idle stop screw
L. Low idle screw

M. Maximum fuel stop screw
S. Shut-off plunger adjusting screw

disconnect shut-off cable, loosen jam nut and back plunger unit out several turns. Operate engine at approximately 900 rpm. Move shut-off lever rearward until engine speed increases. Hold lever in this position, turn plunger unit in until end of plunger just contacts the lever, then tighten jam nut. Move lever fully rearward to depress the spring loaded plunger. Engine must shut off at this position. Connect shut-off cable to control lever.

All Series Equipped With Roosa-Master Pump

178. LOW & HIGH IDLE. To adjust the engine low and high idle speeds, start engine and bring to operating temperature. Disconnect throttle control rod from pump control lever. Hold pump control lever in high idle position (rearward against stop) and check engine rpm. Engine high idle speed should be 2507 rpm for series 706 and 2706, 2630 rpm for series 806, 2806, 1206 and 21206 and 2650 rpm for series 856, 2856, 1256, 21256, 1456 and 21456. Adjust high idle stop screw on rear of pump control lever as necessary to obtain correct high idle rpm.

Release the pump control lever and check engine low idle rpm. Low idle speed should be 675 rpm for series 806, 2806, 1206 and 21206 and 650 rpm for all other series. Adjust engine low idle rpm with the screw on top of pump cover.

With engine operating at low idle rpm, place throttle control lever in low idle position (against dowel pin stop). Adjust length of throttle control rod to hold the pump control lever in low idle position. With control rod connected, place throttle control lever in high idle position. Check to see that pump control lever stop screw is held firmly on its stop by pressure from the control lever spring.

INJECTION NOZZLES

Warning: Fuel leaves the injection nozzles with sufficient force to penetrate the skin. When testing, keep your person clear of the nozzle spray.

All Diesel Models

179. TESTING AND LOCATING FAULTY NOZZLE. If engine does not run properly and a faulty injection nozzle is suspected, or if one cylinder is misfiring, locate the faulty nozzle as follows: Loosen the high pressure line fitting on each nozzle holder in turn, thereby allowing fuel to escape at the union rather than enter the cylinder. As in checking spark plugs in a spark

ignition engine, the faulty nozzle is the one that when its line is loosened least affects the running of the engine.

Remove the suspected nozzle as outlined in paragraph 180, 181 or 182, place nozzle in a test stand and check the nozzle against the following specifications:

Series 706-2706 (D282)

Opening pressure, new ..950-1050 psi
Opening pressure, used ..850 psi min.
Nozzle should not leak at 700 psi for 10 seconds.

Series 706-2706-756-2756 (D310)

Opening pressure, new ......3000 psi
Opening pressure, used ....2900 psi
Nozzle showing visible wetting on the tip after 10 seconds at 2700 psi is permissible. Maximum leakage through return port is 10 drops in 1 minute at 2700 psi.

Series 806-2806-856-2856

Opening preseure, new 2375-2450 psi
Opening pressure, used 2275-2350 psi
Nozzle showing visible wetting on the tip after 5 seconds at 1800 psi is permissible. Maximum leakage through return port is 10 drops in 1 minute at 1500 psi.

Series 1206-21206-1256-21256-1456-21456

Opening pressure, new 3100-3200 psi
Opening pressure, used 2950-3050 psi
Nozzle showing visible wetting on the tip after 5 seconds at 2500 psi is permissible. Maximum leakage through return port is 10 drops in 1 minute at 1500 psi.

Series 706-2706 (D282)

180. **R&R NOZZLES.** To remove any injection nozzle, first remove dirt from nozzle, injection line and cylinder head. Disconnect glow plug wire and injection line from injector and immediately cap openings. Remove cap screws securing nozzle in cylinder head, then remove injection nozzle and dust seal ring. An OTC HC-689 puller or equivalent, can be used in withdrawing a stuck nozzle.

When reinstalling, tighten the nozzle retaining cap screws evenly to a torque of 20-25 ft.-lbs.

Series 706-2706-756-2756 (D310)

181. **R&R NOZZLES.** To remove any injection nozzle, first remove dirt from nozzle injection line, return line and cylinder head. Disconnect leakage return line and high pressure injection line from nozzle and immediately cap or plug all openings. Remove the injector stud nuts and carefully withdraw the injector assembly from cylinder head.

NOTE: It is recommended that cooling system be drained before removing injectors.

It is possible that injector nozzle sleeve may come out with injector and allow coolant to enter engine.

When reinstalling, tighten injector stud nuts evenly and to a torque of 8 ft.-lbs.

Series 806-2806-856-2856-1206-21206-1256-21256-1456-21456

182. **R&R NOZZLES.** To remove the injector nozzles first remove dirt from injection lines, return line and injector nozzles. On turbocharged models, remove turbocharger as outlined in paragraph 191. Then, on all models, unbolt and remove the exhaust manifold. Disconnect and remove the leak-off manifold (return line) assembly, then disconnect the high pressure line from injector. Cap or plug all openings immediately. Remove the injector holddown cap screws and lift injector assembly from cylinder head.

NOTE: It is recommended that cooling system be drained before removing injectors. It is possible that injector nozzle sleeve may come out with injector and allow coolant to enter engine.

When reinstalling, tighten nozzle hold-down cap screws evenly (in 2 ft.-lb. increments) to a torque of 10 ft.-lbs.

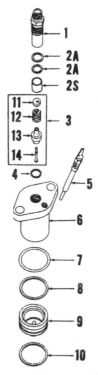

Fig. IH210—Exploded view of IH Midget injector nozzle assembly used on series 706 and 2706 D282 engines.

1. Nozzle fitting
2A. Gaskets
2S. Spacer
3. Valve assembly
4. Gasket
5. Glow plug
6. Body
7. Dust seal
8. Gasket
9. Pre-combustion chamber
10. Gasket
11. Spring seat
12. Valve spring
13. Valve seat
14. Valve (pintle)

Series 706-2706 (D282)

183. **OVERHAUL.** To disassemble the injector, refer to Fig. IH210 and proceed as follows: Remove nozzle fitting (1) from nozzle body, then remove gaskets (2A), spacer (2S), valve assembly (3) and gasket (4). Thoroughly clean and inspect all parts and renew any which are damaged. The gaskets should be renewed each time the nozzle is subjected to complete or partial overhaul.

The nozzle valve assembly (3) is available as a complete unit. To disassemble the nozzle valve for cleaning and/or adjusting the opening pressure, press down on the nozzle spring seat (11) until pressure is relieved from upper end of pintle (14). Then, use a screwdriver to push upper end of pintle to the side releasing the nozzle spring seat. Separate parts and clean in a suitable solvent.

The pintle (14) and seat (13) can be lapped using No. 400 lapping compound.

Nozzle spring seats (11) with flange thicknesses of 0.101-0.102, 0.103-0.104, 0.105-0.106, 0.107-0.108, 0.109-0.110, 0.111-0.112, 0.113-0.114 and 0.115-0.116 are available to adjust the opening pressure. Opening pressure should be adjusted to 950-1050 psi and valve should not leak at seat for 10 seconds at 700 psi.

Nozzle fitting (1) should be tightened in nozzle body (6) to a torque of 45-50 ft.-lbs.

Series 706-2706-756-2756 (D310)

184. **OVERHAUL.** To disassemble the injector, place assembly in a vise with nozzle tip pointing upward. Remove nozzle holder nut (13—Fig. IH-211), then carefully remove nozzle tip (12) and valve (11). Invert the body assembly in vise and remove adjusting lock cap (3). Remove adjusting screw (4), spring seat (5), spring (6), and spindle (7), then remove fuel inlet connector (10).

Thoroughly clean all parts in a suitable solvent. Clean inside the orifice end of nozzle tip with a wooden cleaning stick. The 0.011 diameter orifice spray holes may be cleaned by inserting a cleaning wire of proper size. Cleaning wire should be slightly smaller than spray holes. Clean the fuel return passage in body (8) with a 5/64-inch drill.

When reassembling, make certain all parts are perfectly clean and install parts while wet with clean diesel fuel. To check cleanliness and fit of valve (11) in nozzle tip (12), use a twisting motion and pull valve about ⅓ of its length out of nozzle tip. When released, valve should slide back to its seat by its own weight.

NOTE: Valve and nozzle tip are mated parts and under no circumstance should valves and nozzle tips be interchanged.

Tighten nozzle holder nut (13) to a torque of 50 ft.-lbs. Connect the assembled injector nozzle to a test pump and flush the valve. Adjust opening pressure by turning adjusting screw (4) in to increase or out to decrease opening pressure. Opening pressure should be adjusted to 3000 psi. Valve should not show leakage at orifice spray holes for 10 seconds at 2700 psi.

Series 806-2806-856-2856-1206-21206-1256-21256-1456-21456

185. OVERHAUL. The D361 and DT-361 engines are equipped with either American Bosch or Robert Bosch injector nozzle assemblies shown in Fig. IH212. Robert Bosch injectors are also used on D407 and DT407 engines. Although the construction of the two assemblies are similar, refer to the appropriate following paragraph for overhaul procedure.

186. AMERICAN BOSCH (IH). Place the injector assembly in a vise or holding fixture with the nozzle tip pointing upward. Remove holder nut (13—Fig. IH213) and separate all parts

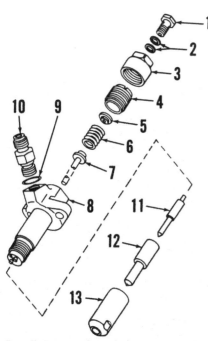

Fig. IH211 — Exploded view of Robert Bosch injector nozzle assembly used on series 706, 2706, 756, 2756 D310 engines.

1. Hollow screw
2. Seal rings
3. Adjusting lock cap
4. Adjusting screw
5. Spring seat
6. Spring
7. Spindle
8. Body
9. Seal ring
10. Fuel inlet connector
11. Valve
12. Nozzle tip
13. Nozzle holder nut

and clean in a suitable solvent. Do not lose shims (5 and 6).

NOTE: On early nozzles, the wear shim (6) is hardened and is a darker color than the other shims. On later nozzles, all the shims are hardened. If servicing an early nozzle and the original shim pack is being reused, be sure the hardened shim contacts the spring (7).

Shims are available in thicknesses of 0.002, 0.003, 0.005 and 0.010.

Inspect mating surfaces of nozzle valve body (12), spring retainer (9) and holder nut (13) for cracks, scratches or other defects. Small defects can be removed by lapping after removing locating roll pins. It is important, however, that sealing faces remain perfectly flat, otherwise a tight seal will not be obtained and leakage will result. The locating roll pins can be removed if necessary by grasping pins with pliers and rotating them in the direction that pin is wound.

Clean inside the orifice end of nozzle (12) with a wooden cleaning stick. Clean the "sac" end of nozzle with a 0.050 inch drill and the fuel return passages with a 5/64-inch drill. Orifice spray holes can be cleaned by inserting a cleaning wire of proper size. Cleaning wire should be slightly smaller than spray holes. Orifice spray hole diameter is 0.011 for D361 engines and 0.013 for DT361 engines. Carbon can be removed from exterior surfaces with a soft brass brush and clean diesel fuel.

When reassembling, be sure all parts are perfectly clean and install parts while wet with clean diesel fuel. To check cleanliness and fit of valve (11) in nozzle, use a twisting motion and pull valve about ⅓ of its length out of the nozzle. When released, valve should slide back to its seat by its own weight.

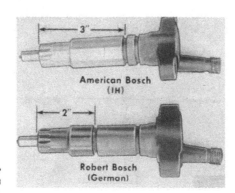

Fig. IH212 — American Bosch (IH) and Robert Bosch injector nozzle assemblies used on D361 and DT361 engines can be identified by the length of the nozzle holder nuts. Robert Bosch injectors are used on D407 and DT4T07 engines.

NOTE: Valve and nozzle (11 and 12) are mated parts and under no circumstance should valve and nozzles be interchanged. Lapping not not recommended on the valve and nozzle seat and faulty units must be renewed.

Reassemble injector by reversing the disassembly procedure. Tighten nozzle holder nut to a torque of 65 ft.-lbs. Adjust opening pressure to 2375-2450 psi on injectors used on D-361 engines and to 3100-3200 psi on injectors used on DT361 engines. Injector should not show leakage at orifice spray holes for 5 seconds at 1800 psi on D361 injectors or for 5 seconds at 2500 psi on DT361 injectors.

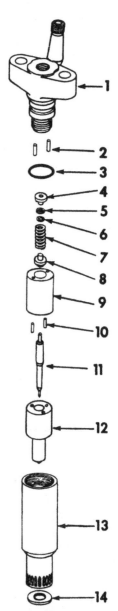

Fig. IH213—Exploded view of American Bosch (IH) injector nozzle assembly.

1. Body
2. Spring pin
3. "O" ring
4. Upper spring seat
5. Adjusting shim
6. Wear shim
7. Spring
8. Lower spring seat
9. Spring retainer
10. Roll pin
11. Valve
12. Nozzle
13. Holder nut
14. Gasket

187. ROBERT BOSCH. Place injector assembly in a vise or holding fixture with the nozzle tip pointing upward. Remove holder nut (8—Fig. IH214), then carefully remove nozzle (7) and valve (6). Remove body (1) from vise, invert the body and remove intermediate plate (5), spring seat (4), spring (3) and adjusting shim (2). Thoroughly clean all parts in a suitable solvent. Inspect mating surfaces of nozzle, intermediate plate and holder nut for cracks, scratches or other defects. Light scratches can be removed by lapping; however, the sealing surfaces must remain perfectly flat. Clean inside the orifice end of nozzle with a wooden cleaning stick. Clean the "sac" end of nozzle with a 0.046 inch drill on injectors used on D361 and D407 engines and a 0.050 inch drill on injectors used on DT361 and DT407 engines. Fuel return passages can be cleaned with a 5/64-inch drill. Orifice spray holes can be cleaned by inserting a cleaning wire of proper size. Cleaning wire should be slightly smaller than spray holes. Orifice spray hole diameter is stamped on nozzle just above orifice tip. Carbon can be removed from exterior surfaces with a soft brass brush and clean diesel fuel.

When reassembling, be sure all parts are perfectly clean and install parts while wet with clean diesel fuel. To check cleanliness and fit of valve (6) in nozzle (7), use a twisting motion and pull valve about 1/3 of its length out of nozzle. When released, valve should slide back to its seat by its own weight.

NOTE: Valve and nozzle (6 and 7) are mated parts and under no circumstance should valves and nozzles be interchanged. Lapping is not recommended on the valve and nozzle seat and faulty units must be renewed.

Reassemble injector by reversing the disassembly procedure. Tighten nozzle holder nut to a torque of 65 ft.-lbs. Adjust opening pressure to 2375-2450 psi for injectors used on D361 and D407 engines and to 3100-3200 psi for injectors used on DT361 and DT-407 engines. Injector should not show leakage at orifice spray holes for 5 seconds at 1800 psi on D361 and D407 injectors or for 5 seconds at 2500 psi on DT361 and DT407 injectors.

PRE-COMBUSTION CHAMBERS

Series 706-2706 (D282 Engine)

188. REMOVE AND REINSTALL. Pre-combustion chambers (9—Fig IH-210) can be pulled from cylinder head after first removing the respective injector nozzle assembly. The use of a special pre-cup puller may be necessary.

When reinstalling, make certain that side stamped "TOP" is toward top of engine and always use new steel gaskets (8 and 10).

GLOW PLUGS

Series 706-2706 (D282 Engine)

189. REMOVE AND REINSTALL. To remove glow plugs (5—Fig. IH-210), disconnect wires at blade connectors, then unscrew and withdraw glow plugs. Glow plugs are not serviceable and faulty units must be renewed.

ETHER STARTING AID

All Diesel Models So Equipped

190. On all diesel series except 706 and 2706 equipped with the D282 engine, it is necessary that ether be used as a starting aid at temperatures below freezing.

To test the ether spray pattern, disconnect ether line at spray nozzle and spray nozzle from manifold air inlet. Reconnect nozzle to ether line. Press ether injection button on dash and observe spray pattern. A good spray pattern is cone-shaped. Dribbling or no spray indicates a blocked spray nozzle or lack of ether pressure. Clean spray nozzle or install new can of ether as needed.

To change the ether fluid container, turn knurled adjusting screw clockwise until container can be removed. Install new container in the bail and tighten adjusting screw (counterclockwise) while guiding container head into position. Rotate container to make certain it is seated properly in injector body, then tighten adjusting screw to hold container firmly in position. CAUTION: Ether must be in twelve ounce containers meeting ICC29 specifications.

NOTE: In warm temperatures, ether container can be removed and a protective plug installed in injector body. DO NOT operate tractor engine without either the ether container or protective plug in position.

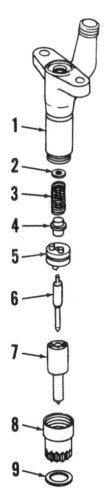

Fig. IH214 — Exploded view of Robert Bosch injector nozzle assembly.

1. Body
2. Adjusting shim
3. Spring
4. Spring seat
5. Intermediate plate
6. Valve
7. Nozzle
8. Holder nut
9. Gasket

DIESEL TURBOCHARGER

Series 1206, 21206, 1256, 21256, 1456 and 21456 diesel engines are equipped with exhaust driven turbochargers. Solar model TC-3B turbocharger is used on series 1206 and 21206 and Schwitzer model 3LD turbocharger is used on series 1256, 21256, 1456 and 21456. The turbochargers consist of the following three main sections. The turbine, bearing housing and compressor.

Engine oil taken directly from the clean oil side of the engine oil filters, is circulated through the bearing housing. This oil lubricates the sleeve type bearings and also acts as a heat barrier between the hot turbine and the compressor. The oil seals used at each end of the shaft are of the piston ring type. When servicing the turbocharger, extreme care must be taken to avoid damaging any of the parts.

NOTE: When the engine has been idle for one month or more, after installation of a new or rebuilt turbocharger or after installing new oil filter elements, the turbocharger must be primed as outlined in paragraph 191.

REMOVE AND REINSTALL

All Series So Equipped

191. To remove the turbocharger, first remove the hood skirts and hood. Unbolt and remove the exhaust elbow, then remove the turbocharger oil inlet and oil drain tubes. Plug or cap all openings. Loosen clamps and disconnect the cross-over tube and the flexible air intake elbow. Unbolt the turbocharger from exhaust manifold, lift the unit from engine and place it in a horizontal position on a bench.

To reinstall the turbocharger, reverse the removal procedure and prime the turbocharger as follows: With the speed control lever in shut-off position, crank engine with starting motor for approximately 30 seconds. Repeat this operation until the engine oil pressure Tellite goes out. Start engine and operate at 1000 rpm for at least 2 minutes before going to a higher speed.

OVERHAUL

All Series So Equipped

192. **SOLAR TURBOCHARGER.** Remove turbocharger as outlined in paragraph 191. Before disassembling the unit, place a row of light punch marks across compressor housing, compressor seal housing, bearing housing and turbine housing to aid in reassembly of the unit. Place unit on bench so that compressor end is facing upward. Remove the four nuts, washers and retainers and remove compressor housing (1—Fig. IH215) from compressor seal housing (6). Unbolt and lift the bearing and compressor seal housings, with rotating assembly and turbine back plate, from the turbine housing. Place the assembly, turbine wheel down, in a mounting stand.

NOTE: A wooden mounting stand made from ¾-inch plywood (See Fig. IH216), twelve inches square, six inches high, open at one end and the bottom, with a five-inch hole cut in the top, should be used to facilitate handling the bearing housing and rotating unit. **Never** rest the weight of the unit on either the turbine or compressor wheel.

Carefully examine the compressor wheel for bent blades, evidence of

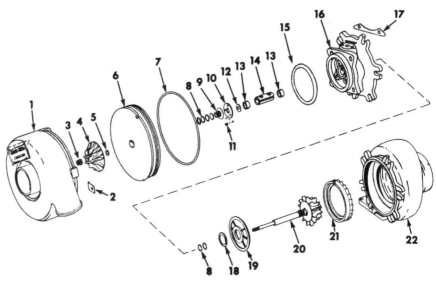

Fig. IH215—Exploded view of Solar model TC-3B turbocharger assembly used on series 1206 and 21206 diesel tractors.

1. Compressor housing
2. Retainer
3. Locknut
4. Compressor wheel
5. Shim (0.002, 0.005 and 0.010)
6. Compressor seal housing
7. "O" ring
8. Seal ring (6 used)
9. Slinger
10. Thrust bearing
11. Self locking screw (3 used)
12. Thrust washer
13. Bearing
14. Spacer
15. Gasket
16. Bearing housing
17. Lock plate
18. Shim (0.005, 0.010 and 0.020)
19. Back plate
20. Turbine shaft and wheel assy.
21. Nozzle ring
22. Turbine housing

rubbing on OD, face and inducer tips of wheel and for pieces of blade broken off. The compressor wheel must be renewed if any of these conditions exist. Then, check the turbine wheel for evidence of rubbing on face of blades or back side of turbine wheel and for bent blades or pieces of blade broken off. Any of these conditions will require the renewal of the turbine wheel. Remove the nozzle ring (21—Fig. IH215) from turbine housing and check for bent or broken vanes. NOTE: Do not attempt to straighten bent blades on the turbine wheel, compressor wheel or nozzle ring.

To remove the compressor wheel, first remove the locknut (3). Then, using a Butane or Oxyacetylene torch, apply heat to the compressor wheel hub until the compressor wheel can be lifted from the turbine shaft. Remove the compressor wheel shims (5).

Withdraw the turbine shaft assembly and turbine bearing from the bearing housing. Remove the turbine back plate, shims and turbine bearing from turbine shaft.

After removing the nuts and washers, remove the compressor seal housing from the bearing housing. The slinger and compressor seal rings can now be removed. Remove the self-locking screws, thrust bearing and thrust washer, then lift compressor bearing and bearing spacer from bearing housing.

Soak all parts in Bendix metal cleaner or equivalent and use a soft brush, plastic blade or compressed air to remove any carbon deposits. CAUTION: Do not use wire brush, steel blade scraper or caustic solution for cleaning, as this will damage turbocharger parts.

Check the turbine shaft bearing surfaces for excessive wear. The shaft journal diameter minimum is 0.437. Inspect the bearings and renew if worn beyond the following limits: Bearing ID is 0.439 maximum and OD is 0.716 minimum. Check the bearing housing for excessive wear in seal ring and bearing bores. The turbine seal ring bore diameter should not exceed 0.725 and the bearing bore should not exceed 0.719. Renew the compressor seal housing if face is

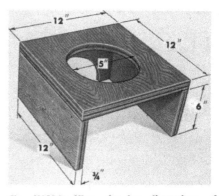

Fig. IH216—View showing dimensions of wooden mounting stand used in disassembly and reassembly of turbocharger.

scored or if seal ring bore is scored or out-of-round. Renew the thrust bearing if worn to less than 0.163 thickness.

Before reassembling the turbocharger, determine the correct turbine back plate shim pack as follows: Place the turbine housing (22—Fig. IH215) in mounting stand and install turbine nozzle ring (21), then place turbine back plate (19), shim (or shims) (18) and bearing housing (16) in position on turbine housing. Using a feeler gage as shown in Fig. IH217, measure gap between the mating surfaces of bearing housing and turbine housing at three equally spaced positions around housings. The average of the three measurements should be 0.003-0.008. If not, select and install correct shim (or shims) (18—Fig. IH215) to obtain this distance. Shims are available in 0.005, 0.010 and 0.020 thicknesses. Separate parts, identify and lay aside the determined shim pack for later reassembly of the turbocharger. This shim pack will provide a 0.003-0.008 interference between outer flange of turbine back plate and the nozzle ring.

When reassembling the turbocharger, use Fig. IH215 and IH218 as a guide and proceed as follows: Place the nozzle ring (21—Fig. IH215) in the turbine housing (22) with the anti-rotating lug in notch provided in housing. Place the turbine shaft and wheel assembly (20) into the turbine housing with the shaft upright. Install two piston type seal rings in grooves on turbine shaft and space the ring gaps approximately 180 degrees apart. Place bearing housing (16), turbine end down, in the mounting stand. Lubricate with engine oil and install the bearing spacer (14), compressor bearing (13), thrust washer (12) and thrust bearing (10). NOTE: Install new thrust bearing if old thrust bearing is worn to less than 0.163 thickness. The heads of the

mounting screws (11) must be flush with or slightly below the face of the thrust bearing. Place bearing housing, compressor end down, in the mounting stand. Install turbine backplate (19) and previously determined shim pack (18) on turbine seal sleeve of bearing housing. Lubricate and install turbine bearing (13) on turbine shaft, then lubricate and install turbine shaft assembly in bearing housing. Lift the bearing housing and turbine shaft assembly from the mounting stand. Hold threaded end of shaft firmly in place and insert turbine wheel into the nozzle in the turbine housing.

Install four seal rings on slinger (9) and space the ring gaps approximately 180 degrees apart. Lubricate and install slinger with seal rings into the compressor seal housing bore. Install compressor seal housing with a new gasket (15), on the bearing housing and after first making certain that punch marks are aligned, secure with nuts and washers. Tighten the nuts evenly to a torque of 4-6 ft-lbs.

Lift the assembly from the turbine housing and place it, turbine end down, in the mounting stand. Place a block under the turbine wheel extension to hold the turbine shaft assembly firmly in place. Using a dial indicator measure the distance of the slinger above or below the face of the compressor seal housing. Add shims (5) to bring the distance to 0.010 above face of the seal housing. Heat the compressor wheel and install it on the turbine shaft. The compressor wheel should be bottomed by hand pressure against the shims. NOTE:

Make certain that balance marks on end of turbine shaft and compressor wheel are in register. Allow the wheel to cool, then install and tighten locknut to a torque of 12-17 ft.-lbs.

With bearing housing in mounting stand, turbine end down, push down on compressor wheel and measure the clearance between compressor wheel and compressor seal housing. Minimum clearance should be 0.010. Then, push rotor assembly upward and again measure the clearance between compressor wheel and compressor seal housing. Maximum clearance must be 0.013-0.020. This will provide the recommended 0.003-0.010 rotor assembly end play. If end play is excessive, disassemble and recheck thrust bearing and thrust washer for excessive wear. Renew parts as needed and reassemble.

Install the assembly in the turbine housing, align the punch marks and bolt the housings together. Tighten bolts evenly to a torque of 5-10 ft.-lbs. Place a new "O" ring (7) in the groove on the outer flange of the compressor seal housing. Spread a small amount of vaseline or equivalent on the "O" ring and mounting flange. Align the punch marks and hand press the compressor housing on the compressor seal housing. Install retainers and tighten the nuts to a torque of 4-6 ft.-lbs.

Check the rotating unit for free rotation within the housings. Cover all openings until the turbocharger is reinstalled.

Use a new gasket and install and prime turbocharger as outlined in paragraph 191.

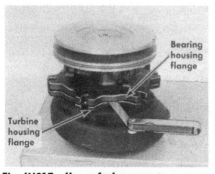

Fig. IH217—Use a feeler gage to measure gap between bearing housing flange and turbine housing flange. Average of three measurements should be 0.003-0.008.

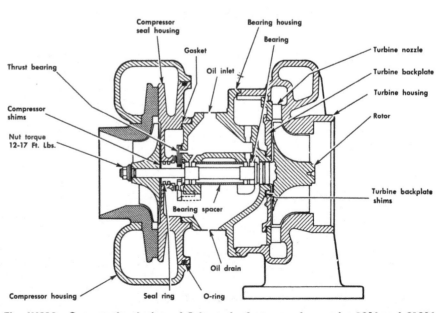

Fig. IH218—Cross-sectional view of Solar turbocharger used on series 1206 and 21206 diesel tractors.

193. SCHWITZER TURBOCHARG-ER. Remove turbocharger as outlined in paragraph 191. Before disassembling, place a row of light punch marks across compressor cover, bearing housing and turbine housing to aid in reassembly. Clamp turbocharger mounting flange (exhaust inlet) in a vise and remove cap screws (14—Fig. IH219), lock washers and clamp plates (13). Remove compressor cover (3). Remove nut from clamp ring (16), expand clamp ring and remove bearing housing assembly from turbine housing (18).

CAUTION: Never allow the weight of the bearing housing assembly to rest on either the turbine or compressor wheel vanes. Lay the bearing housing assembly on a bench so that turbine shaft is horizontal or in a mounting stand similar to that shown in Fig. IH216.

Remove locknut (2—Fig. IH219) and slip compressor wheel (1) from end of shaft. Withdraw turbine wheel and shaft (17) from bearing housing. Place bearing housing on bench with compressor side up. Remove snap ring (7), then using two screwdrivers, lift flinger plate insert (6) from bearing housing. Push spacer sleeve (4) from the insert. Remove oil deflector (11), thrust ring (10), thrust plate (9) and bearing (12). Remove "O" ring (8) from flinger plate insert (6) and remove seal rings (5) from spacer sleeve and turbine shaft.

Soak all parts in Bendix metal cleaner or equivalent and use a soft brush, plastic blade or compressed air to remove carbon deposits. CAUTION: Do not use wire brush, steel scraper or caustic solution for cleaning, as this will damage turbocharger parts.

Inspect turbine wheel and compressor wheel for broken or distorted vanes. DO NOT attempt to straighten bent vanes. Check bearing bore in bearing housing, floating bearing (12) and turbine shaft for excessive wear or scoring. Inspect flinger plate insert, spacer sleeve, oil deflector, thrust ring and thrust plate for excessive wear or other damage.

Renew all damaged parts and use new "O" ring (8) and seal rings (5) when reassembling. The seal ring used on turbine shaft is copper plated and is larger in diameter than the seal ring used on spacer sleve. Refer to Figs. IH219 and IH220 as a guide when reassembling.

Install seal ring on turbine shaft, lubricate seal ring and install turbine wheel and shaft in bearing housing. Lubricate I.D. and O.D. of bearing

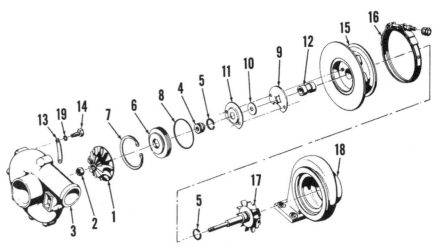

Fig. IH219—Exploded view of Schwitzer model 3LD turbocharger assembly used on series 1256, 21256, 1456 and 21456 diesel tractors.

1. Compressor wheel	7. Snap ring	14. Cap screw
2. Lock nut	8. "O" ring	15. Bearing housing
3. Compressor cover	9. Thrust plate	16. Clamp ring
4. Spacer sleeve	10. Thrust ring	17. Turbine wheel
5. Seal rings	11. Oil deflector	and shaft
6. Flinger plate	12. Bearing	18. Turbine housing
insert	13. Clamp plate	19. Lock washer

(12), install bearing over end of turbine shaft and into bearing housing. Lubricate both sides of thrust plate (9) and install plate (bronze side out) on the aligning dowels. Install thrust ring (10) and oil deflector (11), making certain holes in deflector are positioned over dowel pins. Install new seal ring on spacer sleeve (4), lubricate seal ring and press spacer sleeve into flinger plate insert (6). Position

new "O" ring (8) on insert, lubricate "O" ring and install insert and spacer sleeve assembly in bearing housing, then secure with snap ring (7). Place compressor wheel on turbine shaft, coat threads and back side of nut (2) with "Never-Seez" compound or equivalent, then install and tighten nut to a torque of 13 ft.-lbs. Assemble bearing housing to turbine housing and align punch marks. Install clamp

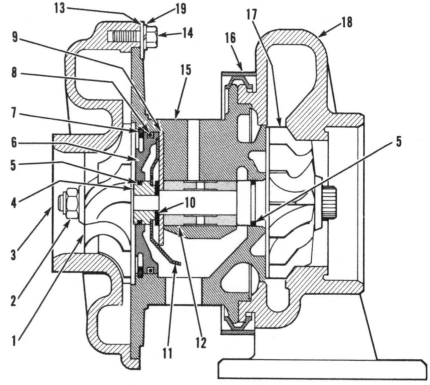

Fig. IH220—Cross-sectional view of Schwitzer turbocharger used on series 1256, 21256, 1456 and 21456 diesel tractors. Refer to Fig. IH219 for legend.

Fig. IH221 — Install test gage as shown to check intake manifold pressure on turbocharged diesel engine. (IH tool numbers are shown).

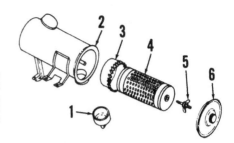

Fig. IH223—Exploded view of typical air cleaner assembly used on series 706, 2706, 806 and 2806 tractors.

1. Dust unloader
2. Housing
3. Air deflector
4. Element
5. Retainer
6. End cover

ring, apply "Never-Seez" on threads and install nut and torque to 10 ft.-lbs. Apply a light coat of "Never-Seez" around machined flange of compressor cover (3). Install compressor cover, align punch marks, and secure cover with cap screws, washers and clamp plates. Tighten cap screws evenly to a torque of 5 ft.-lbs.

Check rotating unit for free rotation within the housings. Cover all openings until the turbocharger is reinstalled.

Use a new gasket and install and prime turbocharger as outlined in paragraph 191.

TESTING

All Series So Equipped

194. Before testing the turbocharger, make certain the air filter system is clean and that the fuel injection pump is properly adjusted and delivering the correct amount of fuel to engine. A clogged air filter will cause intake manifold pressure to be low. Excessive fuel delivery will result in high exhaust manifold pressure and low

fuel delivery will cause low exhaust manifold pressure.

Connect tractor to a dynamometer or other loading device. Remove ¼-inch plug from intake manifold elbow and install a 60 psi test gage as shown in Fig. IH221. Start engine and set speed control lever at high idle position. Load engine to rated load (2400) rpm. Test gage should show intake manifold pressure of 9-12 psi on 1206 and 21206, 8-11 psi on 1256 and 21256 or 10-13 psi on 1456 and 21456. Continue to load engine until overload (1800) rpm is reached. Intake manifold pressure should now be 6-8 psi on 1206 and 21206, 7-10 psi on 1256 and 21256 or 6-9 psi on 1456 and 21456. Stop engine, remove test gage and install plug.

To check exhaust manifold pressure, remove either ¼-inch plug from center of exhaust manifold. Connect a 60 psi test gage, using a length of tubing to protect gage from high temperature. See Fig. IH222. Start engine and set speed control lever at high idle position. Load engine to rated load (2400) rpm. Test gage

should show exhaust manifold pressure of 9-12 psi on 1206 and 21206, 9.5-12.5 psi on 1256 and 21256 or 10-13 psi on 1456 and 21456. Continue to load engine to overload (1800) rpm. Exhaust manifold pressure should now be 5-7 psi on 1206 and 21206, 5.5-8.5 psi on 1256 and 21256 and 4.5-7.5 psi on 1456 and 21456. Stop engine, remove test gage and install plug.

AIR FILTER SYSTEM

All Models

195. All models are equipped with a dry type air cleaner. Series 706, 2706, 806 and 2806 tractors are equipped with the single filter element type shown in Fig. IH223. All other tractors are also equipped with a safety filter element (8—Fig. IH224) which should be renewed at least once each year. DO NOT attempt to clean the safety element.

Large filter element (4—Fig. IH223 or IH224) can be cleaned by directing compressed air up and down the pleats on the inside of the element. Air pressure must not exceed 100 psi. An element cleaning tool (IH tool No. 407073RI) for use with compressed air, is available from International Harvester. Renew filter element after 10 cleanings or once a year, whichever comes first.

Fig. IH222—Test gage installed to check exhaust manifold pressure on turbocharged diesel engine. Length of tubing is used to protect gage from heat.

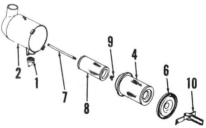

Fig. IH224—Exploded view of typical air cleaner assembly used on all series except 706, 2706, 806 and 2806.

1. Dust unloader
2. Housing
4. Main element
6. End cover
7. Stud
8. Safety element
9. Wing nut
10. Retainer

NON-DIESEL GOVERNOR

The governor used on non-diesel engines, is a centrifugal flyweight type and is driven by the crankshaft gear via an idler gear. Before attempting any governor adjustments, check all linkage for binding or lost motion and correct any undesirable conditions.

ADJUSTMENTS

All Non-Diesel Models

196. **SYNCHRONIZING GOVERNOR AND CARBURETOR.** If removal of carburetor, manifold, governor or governor linkage has been performed, or if difficulty is encountered in adjusting engine speeds, adjust the governor to carburetor control rod as follows: Loosen alternator or generator, remove belt, then move alternator or generator away from engine. Pull throttle lever down to put tension on governor spring, then disconnect control rod from governor (if necessary). Hold both carburetor and governor in the wide open (high idle) position and adjust rod clevis until pin slides freely into clevis and rockshaft lever holes, then remove pin and lengthen rod by unscrewing clevis one full turn. Reinstall clevis pin and tighten jam nut. Install and adjust fan belt.

197. **HIGH IDLE SPEED.** Refer to Fig. IH225 and back out bumper screw about ½ to ¾-inch. Start engine and run until it reaches operating temperature. Move throttle lever to the high idle position and check the engine high idle speed which should be approximately 2530 rpm for the series 706, 2706, 756 and 2756, 2640 rpm for the series 806 and 2806 and 2650 rpm for the series 856 and 2856 tractors. If engine high idle speed is not as stated, loosen jam nut and turn the high idle stop screw as required.

With engine high idle speed adjusted, move the throttle control lever to the low idle position, then quickly advance it to the high idle position and adjust the bumper screw just enough to eliminate engine surge. Repeat this operation as required until engine will advance from low idle to high idle speed without surging. Do not turn bumper screw in further than necessary as engine low idle speed could be affected.

198. **LOW IDLE SPEED.** With engine running and at operating temperature, place throttle control lever in the low idle position and check the engine low idle speed which

should be 425 rpm for all models. If engine low idle speed is not as stated, adjust throttle stop screw on carburetor as required.

If specified engine low idle speed cannot be obtained, recheck the governor to carburetor control adjustment as outlined in paragraph 196. It may also be necessary to vary the length of governor control rod by disconnecting and adjusting the rod ball joint located on rear end of governor control rod.

OVERHAUL

All Non-Diesel Models

199. To remove the governor unit, remove alternator or generator, disconnect governor control rod and governor to carburetor rod from governor, then unbolt governor from engine front plate.

Disassembly of the removed governor unit will be self-evident after an examination of the unit and reference to Fig. IH226.

NOTE: Governor shown is used on series 706 and 2706 tractors. The governor used on other non-diesel series will differ somewhat in that a slightly different thrust sleeve (21) and thrust bearing (18) are used.

The governor weights should have an operating clearance of 0.001-0.010 on the pins. Bearings (18 and 25) should be free of any roughness. The rockshaft lever (9) and governor shaft (27) should operate freely in bearings (11 and 12). Make certain that seal (10) is in good condition. The lubricating holes in governor and engine front end plate must be open and clean.

A governor overhaul service package is available and consists of items 8 through 13 and 15 through 25. A governor gear and weight assembly is also available for service.

After assembly and installation are completed, check and adjust the engine speed as outlined in paragraphs 196, 197 and 198.

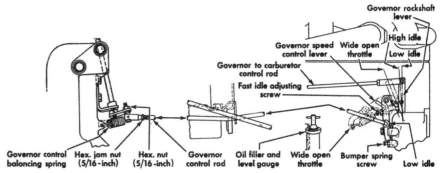

Fig. IH225—Schematic view showing non-diesel engine governor linkage and location of bumper screw and high idle adjusting (stop) screw.

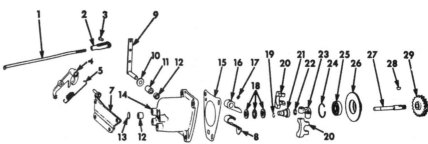

Fig. IH226—Exploded view of typical governor used on non-diesel engines.

1. Governor to carburetor rod	10. Seal	20. Weights
2. Clevis	11. Bearing	21. Thrust sleeve
3. Clevis pin	12. Bearing	22. Snap ring
4. Governor lever	13. Expansion plug	23. Weight carrier
5. Governor spring	14. Housing	24. Snap ring
6.	15. Gasket	25. Bearing
7. Lever bracket	16. Fork	26. Bearing carrier
8. Bumper spring	17. Set screw	27. Governor shaft
9. Rockshaft and lever	18. Thrust bearing	28. Woodruff key
	19. Stop ring	29. Drive gear

COOLING SYSTEM

RADIATOR

All Models

200. To remove radiator, first drain cooling system, then remove hood skirts, hood and side panels. Remove the air cleaner inlet hose. Disconnect upper and lower radiator hoses from radiator and move them out of the way. Disconnect fan shroud from radiator and on all models except 806, 2806, 856 and 2856 non-diesel, remove the radiator drain cock. Disconnect the oil cooler lines bracket. Remove the center cap screw from the four radiator mounts, then lift radiator straight up out of radiator support.

FAN

All Models

201. On series 706 and 2706 equipped with the D282 diesel engine and all non-diesel series, the fan is mounted on a shaft and bearing assembly which is attached by a bracket to the thermostat housing. Two belts are used; one to drive the water pump and alternator or generator and one to drive the fan. To remove the fan and fan shaft assembly, first remove hood skirts and hood. Loosen belt adjuster and remove fan belt. Remove cap screws which retain air cleaner bracket and fan belt adjuster to thermostat housing. Pull fan and fan shaft assembly from side after pushing air cleaner bracket upward out of the way.

Refer to Figs. IH227 and IH228, unbolt and remove fan (23), then press shaft and bearing assembly (19) from pulley (21). Remove retainer (20) and press shaft and bearing from adjuster bracket (18). Shaft and bearing are serviced as an assembly.

When reinstalling, adjust fan belt until a pressure of 25 pounds applied midway between pulleys will deflect belt 7/8-inch.

202. On series 706 and 2706 equipped with the D310 diesel engine and series 756, 2756, 806, 2806, 856, 2856, 1206, 21206, 1256, 21256, 1456 and 21456 diesel tractors, the fan is attached to the water pump and one belt drives the fan, water pump and alternator or generator.

To remove the fan, first remove radiator as outlined in paragraph 200. Remove retaining cap screws and lift off the fan.

When reinstalling, adjust fan belt until a pressure of 25 pounds applied midway between water pump and crankshaft pulleys will deflect belt 7/8-inch.

WATER PUMP

Series 706-2706 (D282 Engine) & All Non-Diesel Series

203. **R&R AND OVERHAUL.** To remove the water pump assembly, drain cooling system, then remove fan assembly as outlined in paragraph 201. Loosen alternator or generator mounting bolts and remove drive belt and pulley. Disconnect by-pass hose and water pump inlet hose. Unbolt and remove water pump assembly.

To disassemble the water pump, refer to Fig. IH227 or IH228 and remove plate (2) and gasket (3). Remove retainer (10) from pump body, support pump body and press shaft and bearing assembly (9) from impeller (4) and pump body. Press shaft assembly from hub (11). Remove seal (6).

The shaft and bearing are available only as a pre-lubricated assembly. Water pump overhaul package (IH part No. 374544R94) is available.

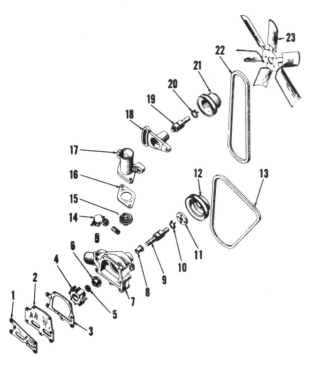

Fig. IH227 — Exploded view of fan shaft and water pump used on all non-diesel engines.

1. Gasket
2. Plate
3. Gasket
4. Impeller
5. Impeller face ring
6. Seal assembly
7. Body
8. Slinger
9. Pump shaft & bearing
10. Retainer
11. Hub
12. Pulley
13. Alternator-generator and pump belt
14. By-pass hose
15. Thermostat
16. Gasket
17. Water outlet
18. Fan belt adjuster
19. Fan bearing & shaft
20. Retainer
21. Fan pulley
22. Fan belt
23. Fan

Fig. IH228 — Exploded view of fan shaft and water pump used on series 706 and 2706 equipped with the D282 diesel engine. Refer to Fig. IH-227 for legend.

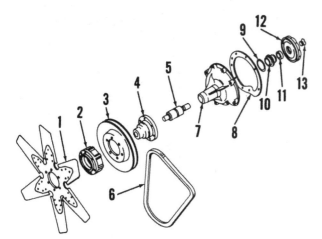

Fig. IH229 — Exploded view of water pump assembly used on series 706, 2706, 756 and 2756 equipped with the D310 diesel engine.

1. Fan
2. Fan spacer
3. Pulley
4. Hub
5. Pump shaft & bearing
6. Alternator & water pump belt
7. Body
8. Gasket
9. "O" ring
10. Seal assembly
11. Impeller face ring
12. Impeller
13. Plastic screw

IH230) need not be removed for water pump overhaul. Therefore, removal and overhaul procedures will be given only for the front (bearing housing) section of the pump.

To remove the water pump, first drain and remove radiator as outlined in paragraph 200. Loosen alternator or generator mounting bolts and remove drive belt. Unbolt and remove fan (2—Fig. IH230) and pulley (5). Unbolt bearing housing (15) from pump rear section (22) and remove the complete front section. Remove Esna nut (4) and using a suitable puller, remove hub (6) and Woodruff key (11). Use a punch on O.D. of seal (8) and collapse seal to remove. Remove bearing snap ring (9). Place pump in a press and force shaft and bearings from impeller (20) and housing (15). Rear bearing (13) may or may not come out with shaft. If bearing remains in housing, it can be pressed out after pump seal (17) is removed. Oil seal (14) will also be removed with the bearing. Front bearing (10) and spacer (7) can also be pressed from shaft (12).

Use all new gaskets and seals when reassembling. Lubricate seals, ceramic washer and "O" ring prior to assembly. Reassemble by reversing the disassembly procedure. Tighten hub retaining Esna nut to a torque of 130 ft.-lbs. Press impeller on shaft until rear face of impeller hub is 2.030-2.070 inches from mounting flange. Remove plug (16) and fill housing (15) with No. 2 multi-purpose lithium grease, then install plug.

Adjust the drive belt as outlined in paragraph 202.

Reassemble by reversing the disassembly procedure, keeping in mind the following points: When installing seal (6), press only on the outer diameter. Install hub with smaller diameter facing out. Press impeller on shaft until a clearance of 0.020-0.030 exists between gasket surface of body and face of impeller.

NOTE: If pump is disassembled for leakage and shaft assembly and impeller are in good condition, renew seal (6) in body and face ring (5) in impeller and reassemble pump.

Reinstall pump by reversing the removal procedure. Adjust water pump and alternator or generator belt tension until a pressure of 25 pounds applied midway between water pump and crankshaft pulleys, will deflect belt ⅝-inch. Install fan and adjust fan belt as in paragraph 201.

Series 706-2706-756-2756 (D310 Engine)

204. **R&R AND OVERHAUL.** To remove the water pump, first remove radiator as outlined in paragraph 200. Loosen alternator mounting bolts and remove the drive belt. Unbolt and remove fan (1—Fig. IH229), spacer (2) and pulley (3). Remove cap screws securing pump body (7) to water pump carrier, then remove pump.

Disassemble water pump as follows: Remove plastic screw (13) and using a ½x2 inch NC cap screw for a jack screw in rear of impeller, force impeller (12) off rear end of shaft. Using two screwdrivers pry seal assembly (10) out of pump body. Support hub (4) and press out shaft. Press shaft and bearing assembly (5) out front of body. Make certain that body is supported as close to bearing as possible.

When reassembling, press shaft and bearing assembly into body using a piece of pipe so pressure is applied

only to outer race of bearing. Bearing race should be flush with front end of body. Install new "O" ring (9) and seal (10). Press only on outer diameter of seal. Support the shaft assembly and press hub on shaft until hub is flush with end of shaft. Install face ring (11) in impeller (12), then press impeller on shaft until there is a clearance of 0.012-0.020 between body and front of impeller (opposite fins). Install plastic screw.

Using a new gasket (8), reinstall pump by reversing the removal procedure. Install pulley, spacer and fan and adjust the belt as outlined in paragraph 202.

Series 806-2806-856-2856-1206-21206-1256-21256-1456-21456 Diesel

205. **R&R AND OVERHAUL.** The water pump is comprised of two sections. The rear section (22—Fig.

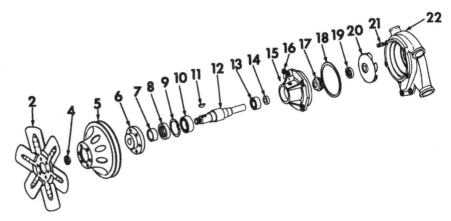

Fig. IH230—Exploded view of water pump assembly used on D361, DT361, D407 and DT407 engines

2. Fan
4. Esna nut
5. Pulley
6. Hub
7. Spacer
8. Oil seal
9. Snap ring
10. Bearing
11. Woodruff key
12. Shaft
13. Bearing
14. Oil seal
15. Bearing housing
16. Plug
17. Pump seal
18. "O" ring
19. Ceramic washer
20. Impeller
21. Stud
22. Rear section

ELECTRICAL SYSTEM

ALTERNATOR AND REGULATOR

All Models So Equipped

206. Delco-Remy "DELCOTRON" alternators and double contact regulators are used on all series except early production 706, 2706, 806 and 2806 tractors.

CAUTION: Because certain components of the alternator can be damaged by procedures that will not affect a D.C. generator, the following precautions MUST be observed:

a. When installing batteries or connecting a booster battery, the negative post of battery must be grounded.

b. Never short across any terminal of the alternator or regulator.

c. Do not attempt to polarize the alternator.

d. Disconnect battery cables before removing or installing any electrical unit.

e. Do not operate alternator on an open circuit and be sure all leads are properly connected before starting engine.

Specification data for the alternators and regulator is as follows:

Alternators 1100687, 1100720 & 1100805

Field current @ 80° F.,
Amperes 2.2-2.6
Volts 12.0
Cold output @ specified voltage,
Specified volts 14.0
Amperes at rpm ...... 21.0 @ 2200
Amperes at rpm .... 30.0 @ 5000
Rated output hot,
Amperes 32.0

Alternator 1100689

Field current @ 80° F.,
Amperes 2.2-2.6
Volts 12.0
Cold output @ specified voltage,
Specified volts 14.0
Amperes at rpm .... 28 @ 2000
Amperes at rpm .... 40 @ 5000
Rated output hot,
Amperes 42.0

Alternator 1100791

Field current @ 80° F.,
Amperes 2.2-2.6
Volts 12.0
Cold output @ specified voltage
Specified volts 14.0
Amperes at rpm .... 32 @ 2000
Amperes at rpm .... 50 @ 5000
Rated output hot,
Amperes 55.0

Regulator 1119516

Ground polarity .... Negative
Field relay,
Air gap 0.015
Point opening 0.030
Closing voltage range 1.5-3.2

Voltage regulator
Air gap (lower points closed) .0.067*
Upper point opening (lower points closed) 0.014
Voltage setting,
65° F. 13.9-15.0
85° F. 13.8-14.8
105° F. 13.7-14.6
125° F. 13.5-14.4
145° F. 13.4-14.2
165° F. 13.2-14.0
185° F. 13.1-13.9

*When bench tested, set air gap at 0.067 as a starting point, then adjust air gap to obtain specified difference between voltage settings of upper and lower contacts. Operation on lower contacts must be 0.05-0.4 volt lower than on upper contacts. Voltage setting may be increased up to 0.3 volt to correct chronic battery undercharging or decreased up to 0.3 volt to correct battery overcharging. Temperature (ambient) is measure ¼-inch away from regulator cover and adjustment should be made only when regulator is at normal operating temperature.

207. **ALTERNATOR TESTING AND OVERHAUL.** The only tests which can be made without removal and disassembly of alternator are the field current draw and output tests. Refer to paragraph 206 for specifications.

To disassemble the alternator, first scribe match marks (M—Fig. IH231) on two frame halves (6 and 16), then remove the four through-bolts. Pry frame apart with a screwdriver between stator frame (11) and drive end frame (6). Stator assembly (11) must remain with slip ring end frame (16) when unit is separated.

NOTE: When frames are separated, brushes will contact rotor shaft at bearing area. Brushes MUST be cleaned of lubricant if they are to be re-used.

Clamp the iron rotor (12) in a protected vise only tight enough to permit loosening of pulley nut (1). Rotor and end frame can be separated after pulley and fan are removed. Check the bearing surfaces of rotor shaft for visible wear or scoring. Examine slip ring surfaces for scoring or wear and windings for overheating or other damage. Check rotor for grounded, shorted or open circuits using an ohmmeter as follows:

Refer to Fig. IH232 and touch the ohmmeter probes to points (1-2) and (1-3); a reading near zero will indicate a ground. Touch ohmmeter probes to the slip rings (2-3); reading should be 4.6-5.5 ohms. A higher reading will indicate an open circuit and a lower reading will indicate a short. If windings are satisfactory, mount rotor in a lathe and check runout at slip rings with a dial indicator. Runout should not exceed 0.002. Slip ring surfaces can be trued if runout is excessive or if surfaces are scored. Finish with 400 grit or finer polishing cloth until scratches or machine marks are removed.

Disconnect the three stator leads and separate stator assembly (11—

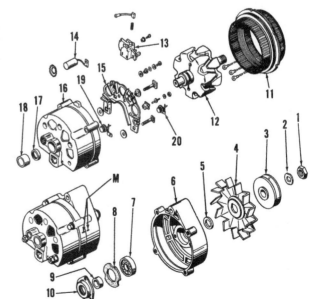

Fig. IH231 — Exploded view of "DELCOTRON" alternator. Note match marks (M) on end frames.

1. Pulley nut
2. Washer
3. Drive pulley
4. Fan
5. Spacer
6. Drive end frame
7. Ball bearing
8. Gasket
9. Spacer
10. Bearing retainer
11. Stator
12. Rotor
13. Brush holder
14. Capacitor
15. Heat sink
16. Slip ring end frame
17. Felt seal and retainer
18. Needle bearing
19. Negative diode (3 used)
20. Positive diode (3 used)

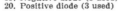

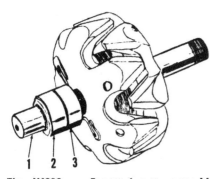

Fig. IH232 — Removed rotor assembly showing test points when checking for grounds, shorts and opens.

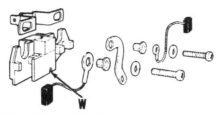

Fig. IH233—Exploded view of brush holder assembly. Insert wire in hole (W) to hold brushes up. Refer to text.

Fig. IH231) from slip ring end frame assembly. Check stator windings for grounded or open circuits as follows: Connect ohmmeter leads successively between each pair of leads. A high reading would indicate an open circuit.

NOTE: The three stator leads have a common connection in the center of the windings. Connect ohmmeter leads between each stator lead and stator frame. A very low leading would indicate a grounded circuit. A short circuit within the stator windings cannot be readily determined by test because of the low resistance of the windings.

Three negative diodes (19) are located in the slip ring end frame (16) and three positive diodes (20) in the heat sink (15). Diodes should test at or near infinity in one direction when tested with an ohmmeter, and at or near zero when meter leads are reversed. Renew any diode with approximately equal meter readings in both directions. Diodes must be removed and installed using an arbor press or vise and suitable tool which contacts only the outer edge of the diode. Do not attempt to drive a faulty diode out of end frame or heat sink as shock may cause damage to the other good diodes. If all diodes are being renewed; make certain the positive diodes (marked with red printing) are installed in the heat sink and negative diodes (marked with black printing are installed in the end frame.

Brushes are available only in an assembly which includes brush holder (13). Brush springs are available for service and should be renewed if heat damage or corrosion is evident. If brushes are re-used, make sure all grease is removed from surface of brushes before unit is reassembled. When reassembling, install brush springs and brushes in holder, push brushes up against spring pressure and insert a short piece of straight

wire through hole (W—Fig. IH233) and through end frame (16—Fig. IH231) to outside. Withdraw the wire after alternator is assembled.

Capacitor (14) connects to the heat sink and is grounded to the end frame. Capacitor protects the diodes from voltage surges.

Remove and inspect ball bearing (7). If bearing is in satisfactory condition, fill bearing ¼-full with Delco-Remy lubricant No. 1960373 and reinstall. Inspect needle bearing (18) in slip ring end frame. This bearing should be renewed if its lubricant supply is exhausted; no attempt should be made to relubricate and re-use the bearing. Press old bearing out towards inside and press new bearing in from outside until bearing is flush with outside of end frame. Saturate felt with SAE 20 oil and install seal and retainer assembly.

Reassemble alternator by reversing the disassembly procedure. Tighten pulley nut to a torque of 45 ft.-lbs.

NOTE: A battery powered test light can be used instead of ohmmeter for all electrical checks except shorts in rotor windings. However, when checking diodes, test light must not be of more than 12.0 volts.

GENERATOR AND REGULATOR

All Models So Equipped

208. Early production 706, 2706, 806 and 2806 tractors were equipped with Delco-Remy D. C. generators and three unit regulators.

Specification data for these units is as follows:

Generators 1100395 & 1100401
Brush spring tension, oz. 28
Field draw,
Volts 12.0
Amperes 1.58-1.67
Cold output,
Volts 14.0
Amperes 20.0
RPM 2300

Generator 1100447
Brush spring tension, oz. 28
Field draw,
Volts 12.0
Amperes 1.58-1.67

Cold output,
Volts 14.0
Amperes 25.0
RPM 3040

Generator 1102337
Brush spring tension, oz. 28
Field draw,
Volts 12.0
Amperes 1.48-1.62
Cold output,
Volts 14.0
Amperes 30.0
RPM 2150

Regulator 1119270E
Ground polarity ....Negative
Cut-out relay,
Air gap 0.020
Point gap 0.020
Closing voltage,
Range 11.8-13.5
Adjust to 12.6
Voltage regulator,
Air gap 0.060
Voltage setting at degrees F.,
14.4-15.4 @ 65°
14.2-15.2 @ 85°
14.0-14.9 @ 105°
13.8-14.7 @ 125°
13.5-14.3 @ 145°
13.1-13.9 @ 165°
Current regulator,
Air gap 0.075
Current setting at degrees F.,
25.0-30.0 @ 65°
24.5-29.0 @ 85°
23.5-28.0 @ 105°
23.0-27.0 @ 125°
21.5-25.5 @ 145°
20.5-24.5 @ 165°
19.5-23.5 @ 185°

STARTING MOTORS

All Models

209. Delco-Remy starting motors are used on all models and specification data for these units is as follows:

Starters 1107275 & 1108334
Volts 12.0
Brush spring tension, oz. 35
No-load test,
Volts 9.0
Amperes (min.) 55.0*
Amperes (max.) 80.0*
RPM (min.) 3500
RPM (max.) 6000
*Includes solenoid

Starters 1113176 & 1113197
Volts 12.0
Brush spring tension, oz. 80
No-load test,
Volts 9.0
Amperes (min.) 50.0*
Amperes (max.) 70.0*
RPM (min.) 3500
RPM (max.) 5500
*Includes solenoid

Starter 1113647

Volts 12.0
Brush spring tension, oz. 80
No load test,
    Volts 9.0
    Amperes (min.) 75.0*
    Amperes (max.) 105.0*
    RPM (min.) 5000
    RPM (max.) 7000
          *Includes solenoid

STARTER SOLENOID

All Models

210. All starting motors are equipped with Delco-Remy solenoid switches. Specification data for these units is as follows:

Solenoid 1114356

Rated voltage 12.0
Current consumption,
    Pull-in winding,
      Volts 5.0
      Amperes 13.0-15.5
    Hold-in winding,
      Volts 10.0
      Amperes 18.0-20.0

Solenoid 1115510

Rated voltage 12.0
Current consumption,
    Pull-in winding
      Volts 5.0
      Amperes 26.0-29.0
    Hold-in winding
      Volts 10.0
      Amperes 18.0-20.0

STANDARD IGNITION

All Non-Diesel Models

211. All non-diesel engines with standard ignition systems are equipped with IH distributors and the firing order is 1-5-3-6-2-4. Overhaul procedure for the distributor is obvious after an examination of the unit and reference to Fig. IH234. Breaker contact gap is 0.020 on all models. Breaker arm spring tension should be 21-25 oz.

212. **DISTRIBUTOR INSTALLATION AND TIMING.** With the oil pump properly installed as outlined in paragraph 82, make certain the timing pointer is aligned with the TDC mark on crankshaft pulley or flywheel.

Install the distributor so that rotor arm is in the number one firing position and adjust breaker contact gap to 0.020. Loosen distributor clamp bolts, turn distributor counter-clockwise until breaker contacts are closed; then rotate distributor clockwise until contacts are just beginning to open. Tighten clamp bolts. Attach a timing light and with engine operating at correct high idle, no-load speed, adjust distributor to the following crank-

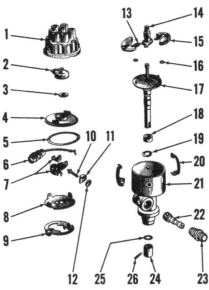

Fig. IH234—Exploded view of IH distributor used on non-diesel engines.

1. Distributor cap	14. Breaker cam
2. Rotor	15. Weight
3. Felt seal	16. Washer
4. Cover	17. Shaft assembly
5. Gasket	18. Oil seal
6. Condenser	19. Thrust washer
7. Breaker contact set	20. Cap retainer
8. Breaker plate	21. Distributor housing
9. Weight guard	22. Tachometer gear
10. Primary terminal	23. Housing
11. Insulator	24. Collar
12. Insulating washer	25. Thrust washer
13. Weight spring	26. Pin

shaft pulley or flywheel degree marks:

Series 706-2706 (C263)
    Gasoline ....22° BTDC
Series 706-2706-756-2756 (C291)
    Gasoline ....18° BTDC
Series 806-2806-856-2856
(C301) Gasoline ....22° BTDC
Series 706-2706-756-2756-806-
  2806-856-2856 LP-Gas ...24° BTDC

MAGNETIC PULSE IGNITION

Non-Diesel Models So Equipped

213. Series 756, 2756, 856 and 2856 non-diesel engines may be equipped

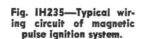

Fig. IH235—Typical wiring circuit of magnetic pulse ignition system.

1. Ignition coil
2. Resistor (0.43 ohm)
3. Pulse amplifier
4. Distributor
5. Resistor (0.68 ohm)
6. Ignition switch
7. Starter switch
8. Starter solenoid switch
9. Battery
10. Connector body

with the "Delcotronic" transistor controlled magnetic pulse ignition system. This system consists of a special pulse distributor, pulse amplifier, resistors and special ignition coil. Refer to Fig. IH235. Ignition switch, starter, solenoid switch and battery are conventional.

The external appearance of the magnetic pulse distributor resembles a standard distributor; however, the internal construction is different. The timer core (4—Fig. IH236) and magnetic pickup assembly (5) are used instead of the conventional breaker plate, breaker contact set and condenser assembly. Timer core (4) has the same number of equally spaced projections as engine cylinders. The magnetic pickup assembly (5) consists of a ceramic permanent magnet, pickup coil and pole piece. The flat metal pole piece has the same number of equally spaced internal projections as the timer core. The timer core which is secured to advance cam (7) is made to rotate around distributor shaft (10) by advance weights (9). This provides a conventional centrifugal advance.

The pulse amplifier (3—Fig. IH235) consists primarily of transistors, resistors, capacitors and a diode mounted on a printed circuit panelboard (10—Fig. IH237).

214. **DISTRIBUTOR INSTALLATION.** With oil pump properly installed as outlined in paragraph 82, make certain timing pointer is aligned with the TDC mark on crankshaft pulley or flywheel. Install distributor so that rotor arm is in number one firing position and projections on pole piece and timer core are aligned. Secure distributor with clamps and bolts.

Attach a timing light and with engine operating at correct high idle, no-load speed, adjust distributor to

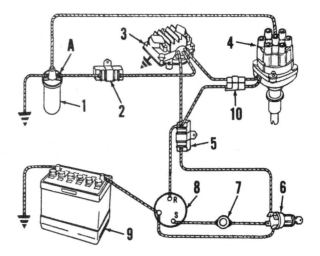

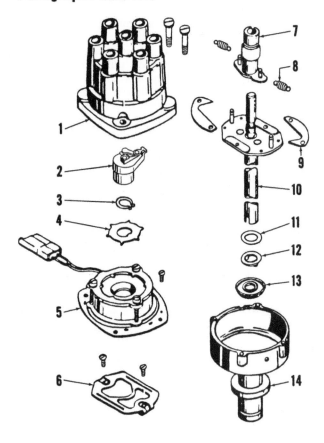

Fig. IH236 — Exploded view of Delco-Remy distributor used on non-diesel engines equipped with magnetic pulse ignition system.

1. Distributor cap
2. Rotor
3. Snap ring
4. Timer core
5. Magnetic pickup assembly
6. Weight hold-down
7. Advance cam
8. Weight spring
9. Weight
10. Shaft assembly
11. Washer
12. Thrust washer
13. Oil seal
14. Housing

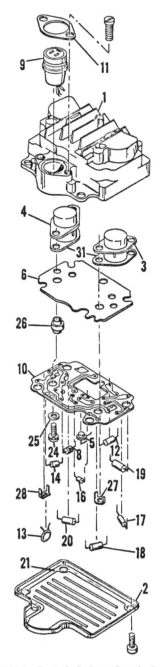

Fig. IH237—Exploded view of typical magnetic pulse amplifier.

1. Housing
2. Cover
3. Drive transistor (TR2)
4. Output transistor (TR1)
5. Trigger transistor (TR3)
6. Heat sink
8. Zener diode
9. Wiring connector
10. Printed circuit panelboard
11. Connector clamp
12. Capacitor
13. Capacitor
14. Resistor (10 ohm)
16. Resistor (680 ohm)
17. Resistor (1800 ohm)
18. Resistor (15000 ohm)
19. Resistor (15 ohm)
20. Resistor (150 ohm)
21. Gasket
24. Screw
25. Washer
26. Bushing insulators
27. Clip (round)
28. Clip (rectangular)
31. Insulators

the following crankshaft pulley or flywheel degree marks:

Series 756-2756

Gasoline ....18° BTDC

Series 856-2856

Gasoline ....22° BTDC

Series 756-2756-856-

2856 LP-Gas ....24° BTDC

215. TROUBLE SHOOTING. Faulty ignition performance usually will be evidenced by one of the following conditions.

A. Engine miss or surge

B. Engine will not run at all.

CAUTION: When trouble shooting the system, use extreme care to avoid accidental shorts or grounds which may cause instant damage to the amplifier. Never disconnect the high voltage lead between coil and distributor and never disconnect more than one spark plug lead unless ignition switch is in "OFF" position. To make compression tests, disconnect wiring harness plug at the pulse amplifier, then remove spark plug leads.

216. If engine misses or surges, and fuel system and governor are satisfactory, check distributor as follows: Make certain that the two distributor leads (solid white and white with green stripe) are connected to connector body as shown in Fig. IH238. Disconnect the connector body halves and connect an ohmmeter (step 1) as shown in Fig. IH238. Any reading above or below a range of 550-750

ohms indicates a defective pickup coil. Remove one ohmmeter lead from connector body and connect to ground (step 2—Fig. IH238). Any reading less than infinite indicates a defective pickup coil. Renew magnetic pickup coil assembly (5—Fig. IH236) if necessary.

A poorly grounded pulse amplifier can also cause an engine to miss or surge. To check, temporarily connect a jumper lead from amplifier housing to a good ground. If engine performance improves, amplifier is poorly grounded. Correct as necessary.

217. If engine will not run at all, remove one spark plug lead and hold end of lead about ¼-inch from engine block. Crank engine and check for spark between spark plug lead and block. If sparking occurs, the trouble most likely is not ignition. If sparking does not occur, check ignition system as follows:

With distributor connector and amplifier connector attached, connect a 12-volt test light to connector body as shown (step 1—Fig. IH239). Turn ignition switch to "ON" position. If test light bulb does not light, there is an open between ignition switch (6—Fig. IH235) and connector body (10), including leads, distributor pickup coil and resistor (5). Check distributor pickup coil as outlined in paragraph 216. If the bulb burns at full brilliance,

resistor (5) is not properly connected to ignition switch. If bulb burns at about half brilliance, resistor (5) is properly connected and circuit is satisfactory. Disconnect ignition wire between switch (6) and resistor (5). Turn ignition switch to "ON" position

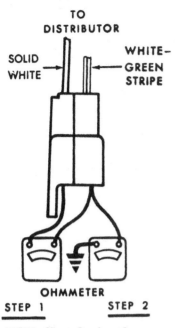

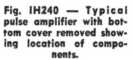

Fig. IH238—View showing ohmmeter connections to distributor connector body when checking for defective pickup coil.

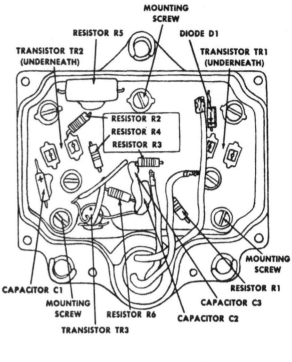

Fig. IH240 — Typical pulse amplifier with bottom cover removed showing location of components.

and press starter switch button (7) to crank engine. If bulb does not light, check for open between solenoid switch (8) and resistor (5). If bulb lights, reconnect ignition wire and proceed with tests. Connect the 12-volt test light between connector body (step 2—Fig. IH239) and coil input terminal (point A—Fig. IH235). Turn ignition switch to "ON" position, press

starter switch button and crank engine. If the bulb flickers, the primary circuit is operating normally. Check (in usual manner for standard ignition system) the secondary system, including spark plugs, wiring, ignition coil tower and secondary winding and the distributor cap for evidence of arc-over or leakage to ground. If bulb does not light, check coil ground wire and coil. If test light burns at full brilliance, check for an open between point "A" and connector body. This includes leads, connections and resistor (2). If wiring, connections and resistor are satisfactory, check for poor amplifier ground by connecting a jumper lead between amplifier housing and a good ground. If bulb now burns at half brilliance, amplifier is poorly grounded.

Correct as required. If bulb remains at full brilliance, renew or repair pulse amplifier as outlined in paragraph 218.

218. AMPLIFIER TEST AND REPAIR. To check the pulse amplifier for defective components, remove the unit from engine and proceed as follows: Refer to Fig. IH237 and remove retaining screws, cover (2) and gasket (21). To aid in reassembly, note location of the three lead connections to the panelboard. See Fig. IH240. Remove three panelboard mounting screws and lift the assembly from amplifier housing.

CAUTION: Drive transistor TR2 (3 —Fig. IH237) and output transistor TR1 (4) are not interchangeable and

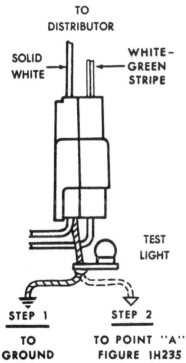

Fig. IH239—View showing test light connections when checking for open circuits or defective pulse amplifier.

Fig. IH241 — View showing test points when using an ohmmeter to check the amplifier components.

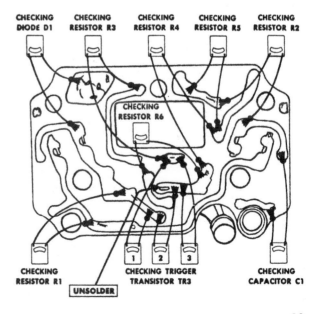

must not be installed in reverse position. Before removing transistors, identify and mark each transistor and their respective installation on the heat sink and panelboard assembly.

Remove the screws securing TR1 and TR2 transistors to panelboard, then separate transistors and heat sink from panelboard. Note the thin insulators (31) between transistors and heat sink and the bushing insulators (26) separating the heat sink from panelboard. Visually inspect the panelboard for defects.

NOTE: To check the panelboard assembly, it is first necessary to unsolder capacitors (C2 and C3—Fig. IH-240) at location shown in Fig. IH241. A 25-watt soldering gun is recommended and 60% tin—40% lead solder should be used when resoldering. DO NOT use acid core solder. Avoid excessive heat which may damage the panelboard. Chip away any epoxy involved and apply new epoxy (Delco-Remy part No. 1966807).

An ohmmeter having a 1½-volt cell is recommended for checking the amplifier components. The low range scale should be used in all tests except where specified otherwise. In all of the following checks, connect ohmmeter leads as shown in Fig. IH241, then reverse the leads to obtain two readings. If, during the following tests, the ohmmeter readings indicate a defective component, renew the defective part, then continue checking the balance of the components.

Trigger Transistor TR3. If both readings in steps 1, 2 or 3 are zero or if both readings in steps 2 or 3 are infinite, renew the transistor.

Didode D1. If both readings are zero or if both readings are infinite, renew the diode.

Capacitor C1. If both readings are zero, renew the capacitor.

Capacitors C2 & C3. Connect ohmmeter across each capacitor. If both readings on either capacitor are zero, renew the capacitor.

Resistor R1. If both readings are infinite, renew the resistor.

Resistor R2. Use an ohmmeter scale on which the 1800 ohm value is within the middle third of the scale. If both readings are infinite, renew the resistor.

Resistor R3. Use an ohmmeter scale on which the 680 ohm value is within the middle third of the scale. If both readings are infinite, renew the resistor.

Resistor R4. Use an ohmmeter scale on which the 15000 ohm value is within the middle third of the scale. If either reading is infinite, renew the resistor.

Resistor R5. Use the lowest range ohmmeter scale. If either reading is infinite, renew the resistor.

Resistor R6. Use an ohmmeter scale on which the 150 ohm value is within the middle third of the scale. If both readings are infinite, renew the resistor.

Transistors TR1 and TR2. Check each transistor as shown in Fig. IH-242. If both readings in steps 1, 2 or 3 are zero or if both readings in steps 2 or 3 are infinite, renew the transistor.

Reassemble by reversing the disassembly procedure. When installing transistors TR1 and TR2, coat tran-

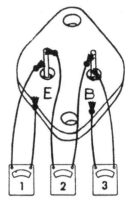

Fig. IH242—Test procedure (steps 1, 2 & 3) when checking TR1 and TR2 transistors with an ohmmeter. Transistors must be installed with emitter pin (E) and base pin (B) in their original positions on panelboard.

sistor side of heat sink (6—Fig. IH237) and both sides of flat insulators (31) with silicone grease. The silicone grease, which is available commercially, conducts heat and thereby provides better cooling.

Delco-Remy part numbers for the pulse amplifier and components parts are as follows:

Pulse amplifier assembly 1115005
Output transistor TR1 1960632
Drive transistor TR2 1960584
Trigger transistor TR3 1960643
Zener diode D1 1960642
Printed circuit panelboard ..1963865
Capacitor C1 1960483
Capacitors C2 and C3 1962104
Resistor R1 (10 ohm) 1960640
Resistor R2 (1800 ohm) 1960639
Resistor R3 (680 ohm) 1960638
Resistor R4 (15000 ohm) 1960641
Resistor R5 (15 ohm) 1963873
Resistor R6 (150 ohm) 1965254

CLUTCH

All Models

219. Series 706, 2706, 756, 2756, 806, 2806, 856 and 2856 tractors are fitted with a 12 inch dry disc clutch. Spring loaded clutches are standard equipment, however, an over-center clutch is optionally available on International models.

Series 1206, 21206, 1256, 21256, 1456 and 21456 tractors are equipped with a 14-inch dry disc spring loaded clutch.

Clutch wear is compensated for by adjusting clutch linkage and when engine clutch is adjusted, it will require that the transmission brake, torque amplifier dump valve (if so equipped) and starter safety switch also be adjusted.

ADJUSTMENT

All Models

220. **ENGINE CLUTCH (SPRING LOADED).** To adjust the spring loaded clutch, refer to Fig. IH243 and disconnect transmission brake rod from brake lever (M) by removing cotter pin (A) and pin (B). Loosen jam nuts (F) and rotate clutch rod turnbuckle (G) as required to obtain ⅞-inch pedal free travel on series 706, 2706, 806, 2806, 1206 and 21206 or ¹¹⁄₁₆-inch pedal free travel on series 756, 2756, 856, 2856, 1256, 21256, 1456 and 21456. Measurement is made between clutch pedal and platform. Tighten jam nuts (F).

Clutch linkage should be adjusted when pedal free travel has decreased to ⅛-inch on series 706, 2706, 806, 2806, 1206 and 21206 or ⅜-inch on all other series.

221. **ENGINE CLUTCH OVER-CENTER).** To adjust the over-center clutch, be sure the ignition or injection system is in "OFF" position. Place clutch lever in the disengaged position and remove the bottom cover (B—Fig. IH244) from clutch housing. Turn engine until adjusting ring lock is on bottom side, then remove lock and turn (tighten) adjusting ring until 30-40 pounds of effort applied to hand lever is required to engage clutch. When properly adjusted, clutch

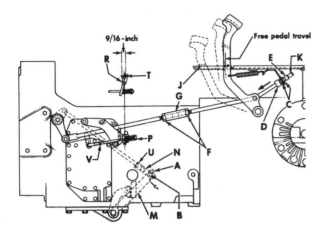

Fig. IH243 — View of spring loaded clutch linkage and points of adjustment. Refer to text for procedure.

A. Cotter pin
B. Clevis pin
C. Lock nuts
D. Safety switch
E. Boot
F. Jam nuts
G. Turnbuckle
J. Pedal stop lug
K. Operating lever
M. Brake lever
N. Clevis
P. Operating screw
R. Operating lever
T. Valve spool pin
U. Jam nut
V. Jam nut

will go over-center with a distinct snap. Install adjusting ring lock and the clutch housing bottom cover. Refer to paragraph 222 for linkage adjustment.

NOTE: When engine is idling, the effort required to engage clutch will be approximately 10 pounds less than when engine is stopped.

222. LINKAGE ADJUSTMENT. Whenever the engine clutch has been adjusted, the control linkage should be checked and adjusted, if necessary.

With engine clutch adjusted, the clutch rod length should be such that clevis pin (C) will freely enter holes of clevis (F) and operating (hand) lever when engine clutch and hand lever are both in the fully disengaged position.

If adjustment of control rod is required, proceed as follows: Engage clutch and loosen jam nut (E). Remove clevis pin (C) and move hand lever to the extreme rear position. This will free clevis so it can be turned and the rod length adjusted as previously stated.

223. TRANSMISSION BRAKE. On tractors with spring loaded clutch, the transmission brake should be adjusted each time engine clutch is adjusted. To adjust the transmission brake, dis-

connect control rod from brake lever (M—Fig. IH243), depress the clutch pedal until stop lug (J) contacts platform, move lever (M) rearward as far as possible, then adjust clevis (N) until clevis pin will freely enter clevis and brake lever. Now remove pin and unscrew clevis ½-turn. Reinstall clevis on brake lever and tighten jam nut.

NOTE: If gear clash is experienced, unscrew clevis one turn additional.

224. DUMP VALVE. The torque amplifier dump valve should be checked and adjusted each time engine clutch is adjusted. Refer to paragraph 225 for information on tractors with spring loaded clutches and to paragraph 226 for tractors with over-center clutches.

225. To position the torque amplifier dump valve on tractors with spring loaded clutches, depress clutch pedal until stop lug strikes platform, loosen jam nut (V—Fig. IH243) and turn operating screw (P) until valve operating lever (R) positions the valve spool pin (T) $\frac{9}{16}$-inch in the extended position as shown.

226. On tractors with over-center clutches, position the torque amplifier dump valve as follows: With clutch lever in disengaged position, loosen set

screw (K—Fig. IH244). Be sure jam nuts (G) are tight and that about ⅛-inch of the pull rod extends beyond the rear jam nut. Pull the dump valve spool out ⅜-inch; then, while holding dump valve in this position, slide lever (J) rearward on clutch rod until it is against the jam nuts (G). Tighten set screw (K) to maintain the adjustment.

227. STARTER SAFETY SWITCH. The starter safety switch prevents the tractor engine from being started except when clutch is in the disengaged position. To adjust the starter safety switch, refer to Figs. IH243 and IH244, and proceed as follows: Loosen the two lock nuts which position switch in the bracket and move the switch, in the direction indicated by the arrow, until upper nut contacts plunger boot. On tractors with spring loaded clutches depress clutch pedal until stop lug on pedal strikes platform. On tractors with over-center clutches, pull lever to extreme rear position. Now position switch so that switch plunger is depressed about ⅛-inch and tighten lock nuts to maintain this position.

REMOVE AND REINSTALL

All Models

228. To remove the engine clutch, it is first necessary to separate (split) engine from clutch housing as outlined in paragraph 229. With engine split from clutch housing, removal of clutch from flywheel is obvious. Refer to paragraph 230 for overhaul data.

TRACTOR SPLIT

All Models

229. To split tractor for clutch service, first drain cooling system and remove front hood, rear hood and the steering support housing (cover). Remove batteries, then disconnect all wiring from engine, unclip harness and lay wiring harness rearward on fuel tank. Disconnect tachometer cable at instrument panel and pull it forward. Disconnect oil pressure switch wire and pull it rearward. Identify the power steering oil cooler hoses and disconnect them at rear.

NOTE: It is essential that the power steering oil cooler hoses be correctly reinstalled as the oil flowing to the oil cooler is maintained at 100 psi by a pressure regulating valve in the multiple control valve and is used as it returns to lubricate the differential before draining back to the main reservoir.

Disconnect the two power steering cylinder lines from control (pilot)

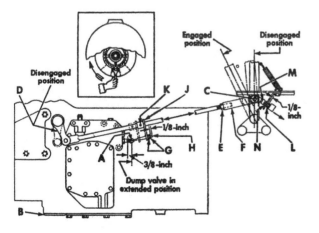

Fig. IH244 — View of overcenter clutch linkage and points of adjustment. Refer to text for procedure.

A. Pin
B. Bottom cover
C. Clevis pin
D. Clutch lever
E. Jam nut
F. Clevis
G. Jam nuts
H. Pull rod
J. Actuating lever
K. Set screw
L. Lock nuts
M. Boot
N. Safety switch

valve. Shut off fuel and disconnect fuel supply line at front end. Disconnect the coolant temperature bulb from cylinder head. Disconnect controls from carburetor, governor or injection pump. On all series except 706, 2706, 756 and 2756, remove the fuel tank left front support and disconnect the tank right front support from tank. Install split stand to side rails and place a rolling floor jack under rear section of tractor. Remove bottom side rail to clutch housing cap screw from both side rails and install guide studs. Complete removal of clutch housing retaining cap screws and separate tractor.

Note:Insert cap screws back in side rails before pulling side rails completely off guide studs.

Rejoin tractor sections by reversing the splitting procedure; however, in order to avoid any difficulty which might arise in trying to align splines of clutch assembly and the transmission input shafts during mating of sections, most mechanics prefer to remove clutch from flywheel and place it over the transmission input shafts. Clutch can be installed on flywheel after tractor sections are joined by working through the opening at bottom of clutch housing.

OVERHAUL

All Models

230. The disassembly and adjusting procedure for either clutch will be obvious after an examination of the unit and reference to Figs. IH245, IH-246 and IH247. Clutch (driven) discs are available as a unit only.

Specification data is as follows:

Series 706-2706-756-2756
(Spring Loaded)
Size, inches 12
Springs:
  Number used 12
  Free length, inches.............$2\frac{11}{16}$
  Lbs. test @ height
    inches ....115-125@$1\frac{13}{16}$
  Minimum allowable ......110@$1\frac{13}{16}$
Lever height, inches...........2.301
Back plate to pressure plate,
  inches 1.020

Series 806-2806-856-2856
(Spring Loaded)
Size, inches 12
Springs:
  Number used 12
  Free length, inches.............$2\frac{11}{16}$
  Lbs. test @ height
    inches ....135-145@$1\frac{13}{16}$
  Minimum allowable ......130@$1\frac{13}{16}$
Lever height, inches...........2.301
Back plate to pressure plate,
  inches 1.020

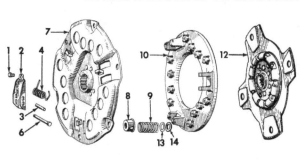

Fig. IH245 — Exploded view of spring loaded clutch used on series 706, 2706, 756, 2756, 806, 2806, 856 and 2856.

1. Lever adjusting screw
2. Lever
3. Pivot pin
4. Lever spring
6. Lever pin
7. Back plate
8. Spring cup
9. Clutch spring
10. Pressure plate
12. Driven disc
13. Insulating washer
14. Washer retainer

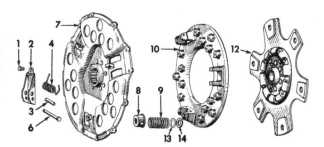

Fig. IH246 — Exploded view of spring loaded clutch used on series 1206, 21206, 1256, 21256, 1456 and 21456. Refer to Fig. IH245 for legend.

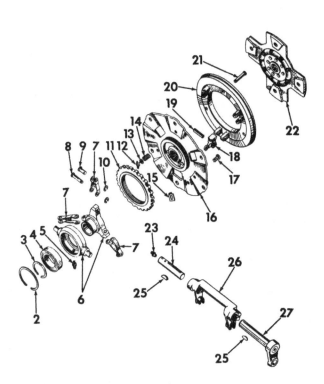

Fig. IH247 — Exploded view showing components of over-center clutch.

2. Snap ring
3. Snap ring
4. Release bearing
5. Grease fitting
6. Release bearing carrier
7. Connecting link
8. Pin (long)
9. Pin (short)
10. Retainer ring
11. Adjusting ring
12. Washer
13. Washer
14. Return spring
    (inner & outer)
15. Adjusting lock
16. Back plate
17. Lock screw
18. Release lever
19. Roll pin
20. Pressure plate
21. Spring stud
22. Driven disc
23. Grease fitting
24. Release shaft (RH)
25. Woodruff key
26. Release fork
27. Release shaft (LH)

Series 1206-21206-1256-21256-1456-
  21456 (Spring Loaded)
Size, inches 14
Springs:
  Number used 12
  Free length, inches ....$2\frac{5}{16}$
  Lbs. test @ height
    inches .... 170 @ $1\frac{21}{32}$
  Minimum allowable .....160 @ $1\frac{21}{32}$
Lever height, inches 2.301
Back plate to pressure plate,
  inches 1.020

CLUTCH SHAFT

All Models

231. The clutch shaft on tractors not equipped with torque amplifier is the transmission input shaft and will be covered in the transmission section.

The clutch shaft on tractors equipped with torque amplifier is part of the torque amplifier assembly and will be covered in the torque amplifier section.

TORQUE AMPLIFIER AND SPEED TRANSMISSION

The torque amplifier and the speed (forward) transmission are both located in the clutch housing along with the hydraulic pump which supplies the power steering, brakes and torque amplifier. Any service on the torque amplifier requires that the entire speed transmission be disassembled before the torque amplifier can be removed. Therefore, this section will concern both units.

Power from the engine is applied directly to the torque amplifier and during operation, the torque amplifier is locked either in direct drive or torque amplifier (underdrive) position by hydraulic multiple disc clutches. There is no neutral position in either the torque amplifier or speed transmission as the neutral position is provided for in the range (rear) transmission.

During operation in the torque amplifier (underdrive) position, the clutch shaft is locked to the torque amplifier constant mesh output gear by a one-way clutch and an approximate 1.48:1 speed reduction occurs, resulting in about a 45 percent increase in torque.

LINKAGE ADJUSTMENT

All Models

232. To adjust the torque amplifier linkage, first remove button plug from control lever, then remove the control lever and the steering support cover. Note the locating punch marks on control lever and shaft. Place control lever back on its shaft (without cap screw) and pull control lever rearward until stop lug on pivot shaft contacts stop pin (J—Fig. IH248). At this time, distance (C) should measure ¾-inch between center of clevis hole and forward edge of clutch mounting flange as shown. If distance is not as stated, disconnect clevis (F) and adjust as necessary. If valve spool prevents adjustment, disconnect valve lever (horizontal) operating rod, then adjust length of bellcrank (vertical) rod until bellcrank is positioned to the ¾-inch measurement.

With bellcrank (D) position determined, reinstall valve lever operating rod and measure distance (W) which should be 1-9/64 inches. If measurement is not as stated, adjust clevis (B) as required.

With the two previous adjustments made, push control lever forward. At this time, the snap ring located on valve spool should be contacting the multiple control valve body and the

stop screw in pivot shaft bracket should be 0.002-0.010 from arm of pivot lever. Adjust stop screw as necessary.

Remove control lever and install steering support cover. Reinstall control lever with aligning punch marks in register and install retaining cap screw and button plug.

To adjust the speed transmission linkage, remove TA control lever and steering support cover. Adjust ball joint linkage at upper end of shift rod to provide accurate positioning of speed transmission lever to the numbers 1-2-3-4 on the quadrant. Shift lever should have clearance at either end of quadrant when transmission is in either first or fourth gear. Reinstall support cover and TA control lever.

REMOVE AND REINSTALL

All Models

233. Removal of the torque amplifier or speed transmission requires the removal of the complete clutch housing from the tractor.

234. To remove clutch housing from tractor, proceed as follows: Remove front hood. Remove torque amplifier control lever and note match marks on lever and shaft. Remove assist handle, steering support cover and center hood. Remove steering support cover skirts. Remove battery or batteries and tray. Disconnect wiring from engine and starter. Remove fuel supply line, and on diesel models, disconnect the fuel return line. Disconnect temperature bulb from engine. On non-diesel models, remove governor rod and disconnect choke control. On diesel models, remove injection pump control rod.

Disconnect the torque amplifier vertical control rod from bellcrank. On series 806, 2806, 856, 2856, 1206, 21206, 1256, 21256, 1456 and 21456 diesel tractors, disconnect tachometer cable at rear end and pull it toward engine. On all other series, disconnect tachometer cable from engine. Disconnect wires from engine oil pressure switch and transmission oil pressure switch. Disconnect power steering hand pump lines from control valve. Identify and disconnect power steering oil cooler hoses at rear end.

NOTE: It is essential that the power steering oil cooler lines be correctly identified for correct installation as the oil flowing to the cooler is controlled at 100 psi by a regulator valve in the multiple control valve. This pressurized oil returning from the oil cooler is used to lubricate the differential gears and bearings.

The pressurized return oil from the power steering valve is used to operate the brakes and torque amplifier clutches as well as providing lubri-

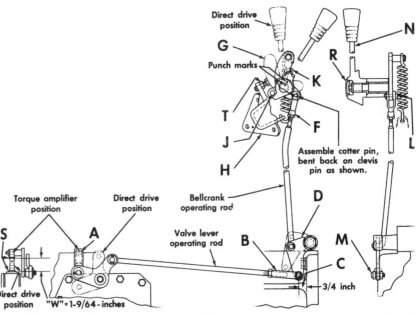

Fig. IH248—Schematic view of torque amplifier operating linkage. Refer to text for adjustment procedure.

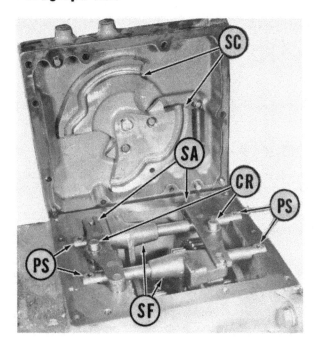

Fig. IH249 — View of speed transmission cover removed.

CR. Cam rollers
PS. Phillips screws
SA. Shifter arms
SC. Shifter cam
SF. Shifter forks

Fig. IH250—Mainshaft, mainshaft bearing assembly and gears shown removed from speed transmission.

OVERHAUL

All Models

235. With the clutch housing removed as outlined in paragraph 234, disassemble the speed transmission and torque amplifier as follows: Disconnect and remove the transmission brake operating rod, then remove snap ring, operating lever and Woodruff key from right end of clutch cross shaft. Loosen the two cap screws in the clutch release fork, bump cross shaft toward left until the two Woodruff keys are exposed and remove Woodruff keys. Complete removal of cross shaft, release fork and throwout bearing assembly.

Unbolt and remove the multiple control valve and pump assembly. Also remove the power steering con-

cation for the torque amplifier clutches and transmission.

On series 706, 2706, 756 and 2756 disconnect tank support from top of clutch housing. On all other series, remove tank left front support and disconnect right front support from tank. On all models, remove platform, drive roll pin from speed transmission shifter cam coupling and disconnect clevis from range transmission shifter arm. Remove control link from park lock. Disconnect the four flexible hydraulic lines at their aft ends, if so equipped. Attach a hoist to steering support and secure tank with a support chain from front of tank to hoist.

NOTE: On International model tractors remove the left and right mud shields from platform, fenders and steering support.

Unbolt steering support from top cover of speed transmission. Carefully lift steering support and tank assembly, disengage coupling from shifter shaft and if necessary, disengage any wiring or tubing which may be interfering.

With tank and steering support removed, disconnect the power steering lines from control valve. Remove the brake control valve line. Remove starter. Remove the torque amplifier control rod, then remove snap ring and the bellcrank and TA valve link. Disconnect clutch rod at both ends, unbolt dump valve pivot bracket and remove rod and bracket assembly. Support front and rear sections of tractor. Unbolt engine and side rails from clutch housing but leave the side rail cap screws in the side rails so they portrude through the engine

rear end plate to provide support. Move sections apart, then attach hoist to clutch housing and separate clutch housing from rear frame.

Note: Spacer on front end of pto shaft may come off as clutch housing is removed. If spacer does drop off, BE SURE to reinstall it during assembly.

Fig. IH251—Speed transmission brake assembly is incorporated in clutch housing bottom cover. Brake pad operates against direct drive constant mesh gear.

Fig. IH252 — View of speed transmission countershaft, gears and spacers.

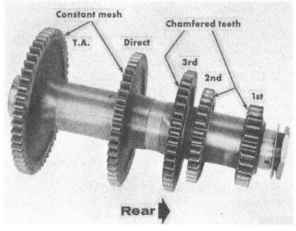

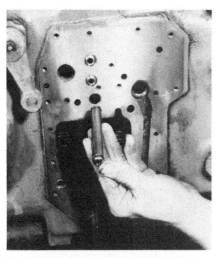

Fig. IH253—Three fluid supply tubes to torque amplifier can be pulled straight out after removal of multiple control valve. Note "O" ring on each end of tube.

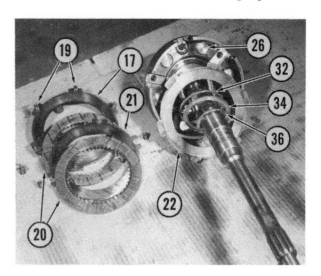

Fig. IH256 — Front (TA lock-up) clutch disassembled. Note the Teflon seal ring and bronze cup washer.

17. Backing plate
19. Springs (6 used)
20. Driven disc
21. Drive plate
22. Backing plate
26. Piston (6 used)
32. Teflon seal
34. Cup washer
36. Thrust washer

Fig. IH254—PTO drive shaft, bearing and bearing cage removed from TA and PTO carrier.

Fig. IH255—The one-way clutch is positioned in inner bore of torque amplifier drive gear.

15. Ball bearing
29. TA drive gear
33. One-way clutch

trol valve. Complete removal of top cover retaining cap screws and lift off the top cover and shifter cam assembly. See Fig. IH249. Lift off shifter arms (SA) and be careful not to lose the two cam rollers (CR). Unstake and remove the four Phillips head screws (PS) which retain the shifter rails and forks, then lift out the shifter rails and forks (SF). Unbolt the transmission main shaft bearing cage and as shaft and bearing cage is pulled rearward, pull gears from top of clutch housing. See Fig. IH250. Remove the clutch housing bottom cover and brake assembly as shown in Fig. IH251. Straighten lock washer on rear end of countershaft and remove the nut. Remove the pto driven gear bearing retainer and the pto driven gear and bearing assembly. Use a thread protector on rear end of countershaft, drive shaft forward and remove gears and spacers from bottom of clutch housing as shaft is moved forward. See Fig. IH252.

Pull the three fluid supply tubes from housing as shown in Fig. IH253.

At this time, the torque amplifier unit can be removed from clutch housing. However, in order to preclude any damage to sealing rings, or other parts, it is recommended that the torque amplifier unit be held together as follows: Install a cap screw and washer in rear end of clutch shaft to hold the direct drive gear assembly in place. Use a "U" bolt around clutch shaft to hold pto drive gear and carrier assembly in place. Install a lifting eye in front end of clutch shaft. Place clutch housing upright with bell end of housing upward. Attach a hoist to the previously installed lifting eye. Unbolt front carrier and the direct drive gear bearing cage (rear) from housing webs and carefully lift torque amplifier assembly from clutch housing.

236. With the torque amplifier assembly removed from clutch housing, disassemble unit as follows: Remove the previously affixed holding fixtures and pull the pto drive gear and carrier assembly and the direct drive gear assembly from the unit. Unbolt and remove the pto drive shaft bearing cage from carrier assembly and remove bearing cage and pto drive shaft as shown in Fig. IH254. Pull torque amplifier drive gear and bearing assembly from TA carrier. Remove lubrication baffle and baffle springs from carrier. Remove seal ring, then remove the one-way clutch from drive gear and be sure to note how the unit is installed in the drive gear. See Figs. IH255 and IH260. It is possible to install the clutch with the wrong side forward and should this happen the torque amplifier would not operate. The quill shaft and output gear can be removed from bearing cage after removing the large internal snap ring. Drive gear and bearing removal from quill gear is obvious. Remove the large internal snap ring from front

Fig. IH257—Rear (direct drive) clutch disassembled. Note inner and outer springs.

37. Piston carrier
43. Driven plate
44. Drive plate
45. Guide pin
46. Inner spring
47. Outer spring
48. Backing plate

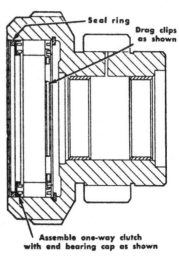

Fig. IH258 — View showing piston removed from piston carrier. Note seal rings on inside and outside diameters of piston. Also note the removed front (lock-up) clutch piston which is installed with widest edge toward front.

25. Clutch carrier
26. Lock-up piston
37. Piston carrier
39. Piston
40. Outer seal
41. Inner seal

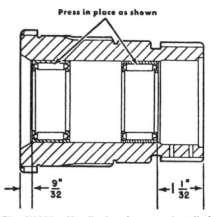

Fig. IH260—Assemble one-way clutch assembly in torque amplifier drive gear as shown. A snap ring and baffle are installed in bore ahead of one-way clutch. Note installation of bearing cap (bronze cup washer) and Teflon seal ring.

(lock-up) clutch and remove the clutch plates as shown in Fig. IH256. Straighten tabs of lock washers at rear of rear clutch, remove nuts and remove direct drive (rear) clutch. See Fig. IH257. Remove the three cap screws and remove the front (lock-up) clutch carrier. Piston carrier is generally a press fit on clutch shaft and can be removed from clutch shaft after removing rear snap ring and thrust washer. Remove piston from piston carrier. See Fig. IH258.

At this time, all parts of the torque amplifier assembly and speed transmission can be inspected and parts renewed as necessary. Procedure for removal of bearings is obvious. Refer to Figs. IH259 through IH262 for installation dimensions and information. Use new "O" rings and lubricate all torque amplifier parts during assembly.

Reassemble components as follows: Install the six pistons in the front (TA) clutch with the widest edge toward clutch discs (front). Install the large piston in the piston carrier with smooth surface toward inside. Align the oil holes of piston carrier with oil holes of clutch shaft and press carrier on clutch shaft. Install thrust washers on each side of clutch carrier with grooved sides away from carrier and install snap rings. Use new gasket, mount front clutch carrier on piston carrier, tighten the three cap screws to 19-21 ft.-lbs. torque and bend down lock washers. Place the nine aligning pins in rear side of piston carrier and place the double springs on pins. Start with back plate next to piston, then install a driven plate (internal spline) and alternate with driving plate (external spline).

NOTE: Early production tractors had direct drive clutches with three 0.070 thick driving plates (external spline) and one 0.070 thick back plate. This was later changed to two 0.100 thick driving plates and one 0.100 thick back plate. In both cases, three driven plates were used. Only the 0.100 thick driving plates and back plate are

available for service and if renewal of these is required, renew as a complete set.

With clutch plates positioned, place the rear clutch backing plate over discs and install the three long bolts but do not tighten nuts until clutch splines have been aligned with the output gear. With clutch discs aligned and the backing plate positioned, tighten nuts to 53-60 ft.-lbs. torque and bend down lock washers. See Fig. IH263 for cross sectional view of assembled clutches. Reassemble the output gear assembly, then install it in the rear clutch and secure it in position with the cap screw and washer used during removal.

Be sure snap ring and baffle is in one-way clutch bore in torque amplifier drive gear, then install the one-way clutch with drag clips toward front as shown in Fig. IH260. Check one-way clutch operation before proceeding further. Slide assembly over hub of piston carrier. Turning gear clockwise should lock the one-way clutch. Counter-clockwise rotation should cause clutch to over-run.

Remove gear and one-way clutch and using heavy grease, position the bronze cup washer over end of one-way clutch, then install the Teflon

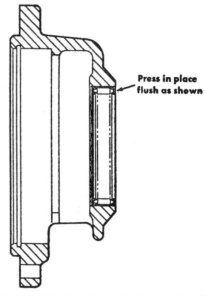

Fig. IH261—Needle bearings are installed in direct drive output quill shaft to dimensions shown.

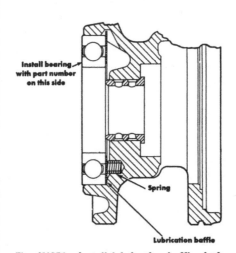

Fig. IH259—Install lubrication baffle, baffle springs and torque amplifier drive gear bearing in TA carrier as shown.

Fig. IH262—Install needle bearing in direct drive output bearing cage as shown.

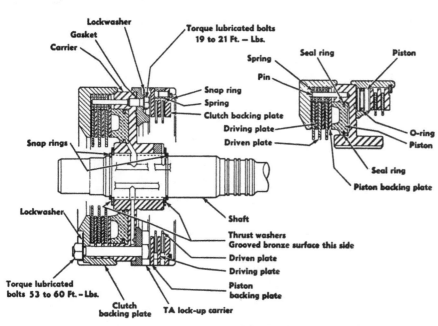

Fig. IH263—Cross-sectional view of TA clutches after assembly.

seal ring. If the large ball bearing is installed on drive gear, place clutch discs over drive gear, install gear and position plate (discs) in clutch carrier. If the bearing is removed from drive gear, install the drive gear and feed clutch discs over drive gear and into clutch carrier. Clutch discs are positioned as follows: Backing plate with no slots in external lugs next to pistons in carrier, driven disc (internal spline), driving disc (with slots in lugs) and driven disc. Place springs through lugs of center driving disc and rest them on lugs of backing plate. Position front backing plate with pins through springs and install the large snap ring.

Renew sealing rings on clutch shaft, install baffle springs and lubrication baffle in TA carrier, then install drive carrier over clutch shaft. If removed, install the pto drive shaft and bearing cage, then install the "U" bolt

1. Bearing cage
2. Oil seal
3. Seal ring
4. Bearing retainer
5. Clutch shaft
6. "O" ring
7. Set screw
8. Carrier
10. "O" ring
11. "O" ring
12. Oil inlet tube
13. Spring (2 used)
14. Lubrication baffle
15. Ball bearing
16. Snap ring
17. Backing plate
18. Dowel pin
19. Clutch spring
20. Driven disc
21. Drive plate
22. Piston backing plate
23. Bolt
24. Lock washer
25. Clutch carrier
26. Piston (TA lock-up)
27. "O" ring
28. Gasket
29. Drive gear
31. Snap ring
32. Teflon seal
33. One-way clutch
34. Baffle
35. Snap ring
36. Thrust washer
37. Piston carrier
38. Steel ball
39. Piston
40. Seal, outer
41. Seal, inner
42. Backing plate
43. Driven disc
44. Drive disc
45. Guide pin
46. Clutch spring, inner
47. Clutch spring, outer
48. Backing plate
49. Lock washer
50. Needle bearing
51. Needle bearing
52. Bearing cage
53. Drive gear
54. Snap ring
55. Ball bearing
56. Quill shaft
57. Snap ring
58. Lock plate

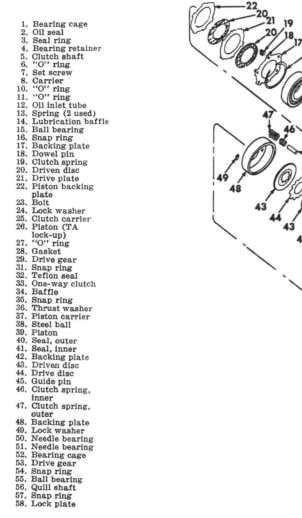
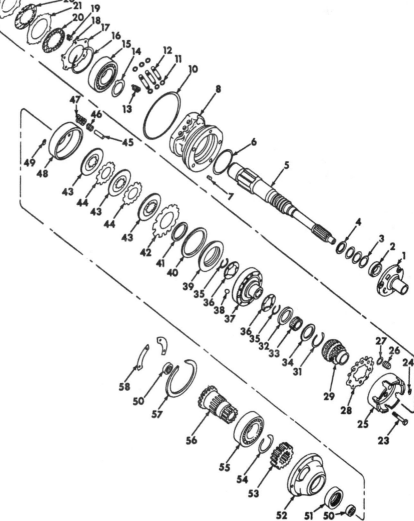

Fig. IH264—Exploded view of torque amplifier assembly showing component parts and their relative positions.

clamp previously used to hold parts in position. Install lifting eye in front end of clutch shaft.

Attach torque amplifier unit to a hoist and lower assembly into clutch housing making sure the cut out portion in direct drive gear bearing cage is on bottom side. Install cap screws in both end bearing cages and tighten securely. Install the three oil inlet tubes. Start speed transmission lower shaft in front of clutch housing and install gears and spacers in the sequence shown in Fig. IH252. Reading from left to right, chamfered teeth of number three gear face front, number four face rear and number five face front. Tighten nut on rear of shaft to 90-100 ft.-lbs. torque. Shaft should rotate freely with no visible end play. If above conditions are not met, recheck installation.

If mainshaft bearing assembly was disassembled, reassemble as follows: Press forward bearing cup into bearing cage with largest diameter rearward. Place bearing cones in the cage and install rear bearing cup. Place end plate, original shims and bearing assembly over mainshaft. Press both bearings on shaft and install the snap ring. Start shaft in rear of clutch housing and place gears on shaft as shaft is moved forward. Install bearing retainer cap screws and rotate shaft while tightening cap screws. Shaft should turn freely with no visible end play. Vary shims as necessary.

Disassembly and reassembly of transmission top cover and shifter cams is obvious after an examination of the unit.

Balance of reassembly is the reverse of disassembly. Shifter forks are interchangeable. Be sure to stake the shift rail retaining screws after installation.

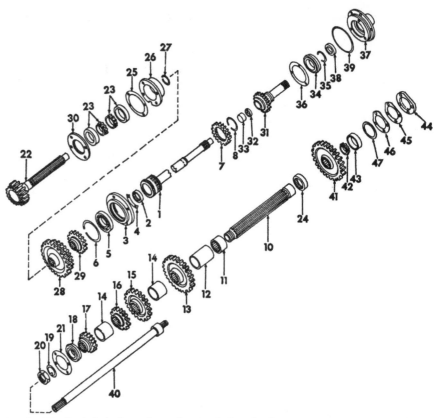

Fig. IH265—Exploded view of speed transmission showing components used when tractor is not equipped with torque amplifier. On early production tractors spacer (11) is used instead of torque amplifier constant mesh gear. Late production tractors use a snap ring on shaft (10) instead of spacer.

1. Transmission drive shaft	16. 2nd speed gear	31. Pto drive shaft
2. Bearing	17. 1st speed gear	32. Oil seal
3. Bearing cage	18. Ball bearing	33. Bushing
4. Lock plate	20. Nut	34. Ball bearing
5. Ball bearing	21. Bearing retainer	35. Snap ring
6. Snap ring	22. Mainshaft	36. Bearing retainer
7. Constant mesh drive gear	23. Bearing assembly	37. Bearing cage
8. Snap ring	24. Needle bearing	38. Oil seal
10. Countershaft	25. Shim (0.007)	39. "O" ring
11. Spacer (short)	26. Bearing cage	40. Pto driven shaft
12. Spacer (long)	27. Snap ring	41. Pto driven gear
13. Constant mesh gear	28. 1st & 2nd sliding gear	42. Bearing cone
14. Spacer	29. 3rd & 4th sliding gear	43. Bearing cup
15. 3rd speed gear	30. Bearing retainer	44. Bearing retainer
		45. Shim, heavy
		46. Shim
		47. Seal

Reinstall the clutch housing assembly by reversing the removal procedure. Adjust the clutch, TA dump valve and transmission brake linkage as outlined in paragraphs 220 thru 227 and the torque amplifier linkage and speed transmission linkage as outlined in paragraph 232.

RANGE TRANSMISSION

The range transmission is located in the front portion of the tractor rear frame. This transmission provides four positions: Hi (direct drive), Lo (underdrive), neutral and reverse. The Lo-range position provides a 3½ to 1 speed reduction and the reverse speed is about 25 percent faster than the same forward speed in the Lo-range. The transmission parking lock is also located in the range transmission.

To remove the range transmission gears and shafts, the tractor rear main frame must be separated from the clutch housing and the differential assembly removed. See Fig. IH266 for an exploded view of the range transmission shafts and gears. Refer to paragraph 237 for information on overhaul of the range transmission and to paragraphs 242 and 243 for linkage adjustment procedures.

R&R AND OVERHAUL

All Models

237. If proper split stands are available, the rear main frame can be separated from the clutch housing before any disassembly is done. However, if range transmission is to be disassembled, it is usually as satisfactory to remove the pto unit, hydraulic lift assembly and the differential before main frame is separated from clutch housing. Thus the main frame can be easily supported for the preliminary disassembly and after the above mentioned components are removed, the main frame can be removed and placed on a bench for service.

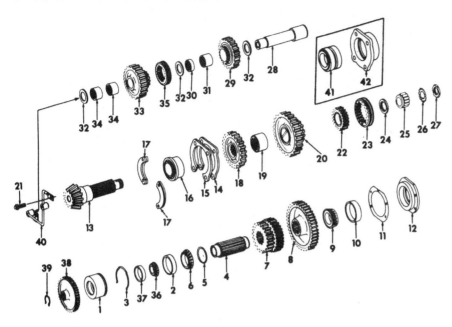

Fig. IH266—Exploded view of range transmission shafts and gears. Bearing assembly (41) and bearing cage (42) shown in inset are used on series 706, 2706, 756, 2756, 806, 2806, 856 and 2856. All other series use bearing assembly (16) and retainers (17).

1. Bearing cage	12. Bearing cage	22. Collar carrier	32. Thrust washer
2. Bearing cup	13. Bevel pinion shaft	23. Shift collar	33. Reverse drive gear
3. Retainer	14. Shim (0.004)	24. Spacer	34. Needle bearing
4. Countershaft	15. Shim (0.007)	25. Roller bearing	35. Shift collar
5. Snap ring	16. Bearing assembly	26. Lock washer	36. PTO bearing cone
6. Bearing cone	17. Bearing retainers	27. Nut	37. Bearing cup
7. Lo-drive gear	18. Reverse driven	28. Reverse idler	38. Hydraulic pump
8. Constant mesh	gear	shaft	drive gear
gear	19. Gear carrier	29. Reverse idler gear	39. Retainer
9. Bearing cone	20. Hi-Lo driven gear	30. Needle bearing	40. Lubrication tube
10. Bearing cup	21. Cap screw	31. Needle bearing	41. Bearing assembly
11. Shim			42. Bearing cage

Fig. IH267 — Right final drive assembly removed from rear frame of Farmall 806. Other series are similar. Use guide pins when reinstalling.

Fig. IH268 — When removing PTO unit, be sure to tip front end upward to prevent damage to oil inlet tube and screen.

238. To disassemble the range transmission, proceed as follows: Drain both sections of the rear frame. Remove platform and seat. Support tractor under clutch housing and place a rolling floor jack under rear frame. Remove rear wheels and both fenders. Remove both brake housings and brake assemblies. Remove all brake lines, then unbolt brake control valve from its mounting, swing valve forward, then unbolt mounting bracket from right final drive housing and remove the valve, bracket and pedals as an assembly. Attach hoist to final drive assemblies and remove final drives from rear frame. See Fig. IH267. Disconnect actuating link from the pto control valve, attach hoist to pto unit, then unbolt and remove the pto unit and extension shaft.

NOTE: As pto unit is withdrawn, be sure to tip front of unit upward as shown in Fig. IH268, otherwise damage to the oil inlet tube and screen will result. Also, be sure to leave the two short cap screws in place so unit will not separate during removal.

Remove the hitch upper link and disconnect lift links from rockshaft lift arms. Disconnect the flexible hydraulic lines at front of lift housing if tractor is so equipped. Disconnect the rear break-away coupling bracket from rear frame. Unbolt hydraulic lift housing from rear frame, attach hoist and lift unit from rear frame. Attach hoist to differential, remove the differential bearing retainers and lift differential from rear frame. Note that bearing on the bevel gear side of differential is larger than the opposite side so be sure to keep bearing retainers in the proper relationship.

NOTE: At this time, most mechanics prefer to remove drawbar, or hitch lower links, to facilitate handling of the rear frame.

Disconnect spring and remove the park lock operating rod (turnbuckle assembly). Disconnect clevis from range transmission shifter arm or arms. Attach hoist to rear frame, unbolt from clutch housing and move rearward until pto drive shaft clears the clutch housing. If the spacer which fits over forward end of pto drive shaft has come off, remove the front bottom cover from clutch housing and retrieve it for installation during reassembly.

239. The range transmission can now be disassembled as follows: Lift park lock shaft about ½-inch and support in this position. Remove top cover cap screws, lift top cover slightly from rear frame and work reverse fork

shaft (10—Fig. IH269 or IH270) out of its bore as top cover is removed Remove nut from front end of mainshaft and using a split collar and puller, remove and discard the mainshaft front bearing. Remove the lubrication tube at rear of mainshaft, then complete removal of the mainshaft bearing retaining cap screws, pull the mainshaft rearward and lift gears from top of housing. Remove cover from left front of rear main frame, pull the reverse shift fork shaft from fork, then turn fork slightly (top inward) and lift from groove of shift collar. Screw a slide hammer in the threaded hole in front of reverse idler shaft, bump shaft forward and remove thrust washers, gears and shift collar from side of main frame. Remove the small retainer at rear of the hydraulic pump drive gear, then attach slide hammer to front of pto drive shaft and

bump shaft forward and out of transmission countershaft. Remove the large retainer at front of countershaft rear bearing cage and remove the bearing cage and bearings. Remove countershaft front bearing retainer and bump shaft rearward. Remove gears from top of main frame.

240. Clean and inspect all parts and renew any which show excessive wear or damage. Refer to Figs. IH271 and IH272 when renewing needle bearings in the reverse idler and reverse drive gears. The mainshaft rear bearing assembly is available only as a package of mated parts which will provide the correct operating clearance. On series 706, 2706, 756, 2756, 806, 2806, 856 and 2856, package contains both bearing cups, both bearing cones and the center spacer. If the mainshaft rear bearing assembly is renewed, proceed as follows: Press both bearing cups into

bearing retainer, with smallest diameters toward center, until they bottom. Press rear bearing on mainshaft, with largest diameter toward gear, until it bottoms. Place bearing cage over mainshaft with flange toward gear (rear), then place spacer over shaft. Press front bearing on shaft with largest diameter toward front and as bearing cone enters bearing cage, rotate the cage to insure alignment of parts. On series 1206, 21206, 1256, 21256, 1456 and 21456, press complete bearing assembly (16—Fig. IH266) on mainshaft

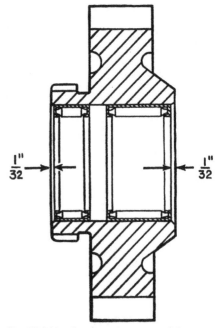

Fig. IH271—Bearings in reverse idler gear are installed as shown.

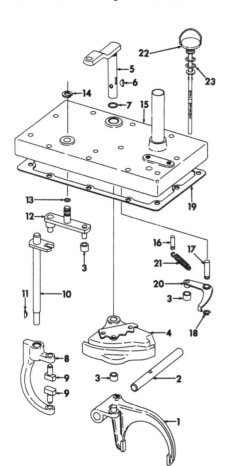

Fig. IH269—Exploded view of range transmission top cover and shifting mechanism used on series 706, 2706, 806, 2806, 1206 and 21206.

1. Hi-Lo shift fork	13. "O" ring
2. Shifter shaft	14. Snap ring
3. Cam roller	15. Cover
4. Shift cam	16. Anchor pin
5. Cam pivot shaft	17. Pivot pin
6. Woodruff key	18. Snap ring
7. "O" ring	19. Gasket
8. Reverse shift fork	20. Control arm
9. Fork pads	21. Spring
10. Reverse fork shaft	22. Oil level gage
11. Woodruff key	23. "O" ring
12. Shift pivot arm	

Fig. IH270—Exploded view of range transmission top cover and shifting mechanism used on series 756, 2756, 856, 2856, 1256, 21256, 1456 and 21456.

1. Hi-Lo shift fork	11. Woodruff key
2. Shifter shaft	12. Reverse pivot shaft
3. Detent roller	
4. Hi-Lo pivot arm	13. Oil level gage
5. Hi-Lo pivot shaft	14. Snap ring
6. Reverse detent sector	15. Cover
	16. "O" ring
7. "O" ring	17. Pivot pin
8. Reverse shift fork	18. Spring
9. Fork pads	19. Gasket
10. Reverse fork shaft	20. Control arm

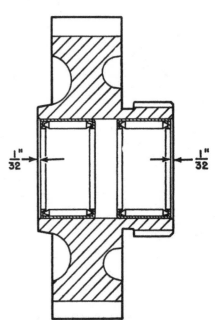

Fig. IH272—Bearings in reverse drive gear are installed as shown.

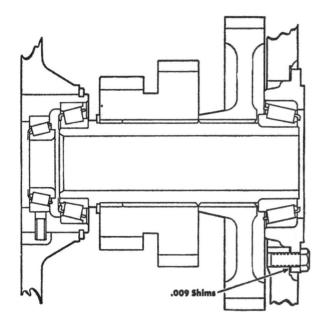

Fig. IH273 — Cross-sectional view showing installation of range transmission countershaft.

.009 Shims

so that flange on outer diameter is to the rear.

Any disassembly or service required on transmission top cover assembly will be obvious after an examination of the unit and reference to Figs. IH269, IH270 and IH276.

241. Reassembly sequence of the range transmission is countershaft, reverse idler shaft, mainshaft and pto drive shaft.

To install countershaft, install the snap ring (5—Fig. IH266) on rear of shaft, then press on bearing cone (6) with largest diameter forward. Note that rear bearing cone is narrower than front bearing cone. Start shaft into housing at rear and install Lo-drive gear (7) with smallest gear rearward and the constant mesh gear (8) with hub rearward on the shaft. Obtain a bolt 8 inches long and two large washers and with bolt inserted through countershaft, pull the front bearing cone on front of countershaft. Large diameter of bearing cone is toward rear. Press both bearing cups (37 and 2) into rear bearing cage (1) with large diameters toward front. Place the small bearing cone (36) in cage, then install cage and secure it with the horseshoe retainer (3). Install bearing cup (10) in bearing cage (12) with large diameter rearward. Place the original shims (11) on bearing cage, or if new parts were installed, use one shim as a starting point. Shims are 0.009 thick. Install bearing cage and tighten cap screws finger tight; then, rotate shaft and tighten cap screws evenly until shaft binds or bearing cage bottoms. If shaft binds, add shims as required, or if shaft has excessive end play, re-

move shims. Shaft should turn freely with no end play. Tighten cap screws to 35 ft.-lbs. torque. See Fig. IH273 for an assembly view.

To install reverse idler shaft and gears, start shaft (28—Fig. IH266) in its bore in front of main frame. As shaft is moved rearward install thrust washer (32), idler gear (29) with shift collar teeth rearward, shift collar (35) and center thrust washer (32), reverse drive gear (33) with shift collar teeth forward and the rear thrust washer (32). Align dowel of shaft with notch in rear frame and bump shaft into position.

NOTE: As shaft is bumped rearward BE SURE the rear thrust washer does not catch the rear step of shaft or thrust washer will be damaged. When properly installed, front

Fig. IH274—Whenever tractor is split between rear main frame and clutch housing, renew oil suction tube "O" ring and retainer.

O. "O" ring
R. Retainer
S. Spacer
25. Bearing

27. Nut
28. Reverse idler shaft
40. PTO shaft

of shaft must be flush with front surface of rear frame. Clutch housing rear surface retains shaft in position.

To install the mainshaft, install the rear bearing assembly as outlined in paragraph 240. Place the original shims (14 and 15) on bearing cage (41) or bearing assembly (16) and start mainshaft in rear frame. Place reverse driven gear on shaft. Place carrier (19) on shaft, then install Hi-Lo driven gear (20) on carrier with shift collar teeth forward. Place shift collar (23) and collar carrier (22) on shaft. Install spacer (24) and a new bearing (25). Install lock washer (26) and nut (27) and tighten nut to a torque of 100 ft.-lbs. Do not install the lubrication tube at this time. Install and tighten mainshaft rear bearing retaining cap screws to a torque of 85 ft.-lbs. Check rotation of shaft. Shaft should turn freely and if binding exists, recheck rear bearing assembly.

Install pinion shaft (mainshaft) locating tool (IH tool No. FES 68-13) as shown in Figs. IH277 and IH278. A cone setting number is etched on rear end of pinion shaft. NOTE: Early production pinion shafts were marked with the etched setting numbers 4 thru 22. Late production pinion shafts are marked with etched setting numbers 60 thru 90. Use this setting number and refer to chart in Fig. IH279 to determine the correct feeler gap (A—Fig. IH278). Add or remove

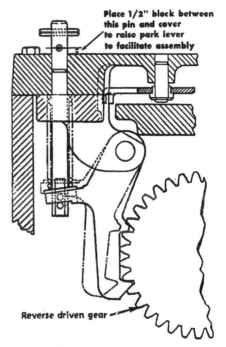

Place 1/2" block between this pin and cover to raise park lever to facilitate assembly

Reverse driven gear

Fig. IH275—When installing range transmission top cover, use a ½-inch spacer positioned as shown to hold park lock in disengaged position.

shims at pinion shaft rear bearing to obtain the determined gap (A) plus or minus 0.001. For example: If a number 9 or 73 is etched on pinion shaft, feeler gap (A) would be 0.045 plus or minus 0.001. Shims are available in thicknesses of 0.004 and 0.007.

After pinion shaft (mainshaft) has been adjusted, remove locating tool and install the lubrication tube (40—Fig. IH266).

Insert the pto drive shaft through the countershaft and into its rear bearing cone.

NOTE: Although pto shaft can be bumped through the rear bearing, it is recommended that a cap screw, washer and a pipe spacer of proper diameter be used to pull the shaft through bearing cone.

Install hydraulic pump drive gear (38) and the small horseshoe retain-

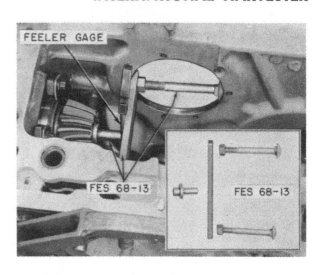

Fig. IH277—Use IH tool No. 68-13 and feeler gage when checking pinion shaft setting.

er (39) on rear of pto shaft.

Install pads in reverse shifter fork, position shifter fork in groove of reverse shift collar, then install the fork shaft and Woodruff key in fork.

Reinstall differential assembly in rear main frame and check carrier bearing preload as outlined in paragraph 244 and backlash as outlined in paragraph 247.

Place spacer on forward end of pto shaft, renew hydraulic pump suction tube "O" ring and retainer (Fig. IH274) and join rear main frame to clutch housing.

Complete reassembly of tractor by reversing the disassembly procedure. Installation of the range transmission top cover will be simplified if a ½-inch spacer is positioned between lower pin of park lock shaft and top cover as shown in Fig. IH275.

LINKAGE ADJUSTMENT

Series 706-2706-806-2806-1206-21206

242. To adjust the range transmission and park lock linkage, first remove the TA control lever and the steering support top and rear covers. Refer to Fig. IH280 and move range

transmission control lever to neutral position. Disconnect ball joint linkage at stud (A) and remove transmission lower linkage pins (B). Remove lower linkage (C). With park lock operating lever in disengaged (up) position, remove pin (G) and loosen jam nut (H). Hold park lock lever about ¾-inch from support housing as shown and with bellcrank (F) in disengaged stop position (against support base) adjust clevis (R) until pin (G) can be installed. Secure pin (G) with cotter pin and tighten nut (H).

Move park lock lever to engaged (down) position so that pawl (J) passes through neutral notch in plate (K). With lever (L) in neutral position and pawl (J) centered in notch of plate (K), adjust clevis (E) on linkage (C) until pins (B) can easily be installed. Secure pins (B) and tighten nut (D).

With range transmission control lever in neutral position, adjust ball joint (upper) linkage until stud (A) will freely enter the hole in lever on vertical shaft. Tighten nut (S) and jam nut (M) securely.

With park lock operating lever in engaged position, disconnect the actuating spring. Back off jam nuts (N & O). Make certain that bellcrank (F)

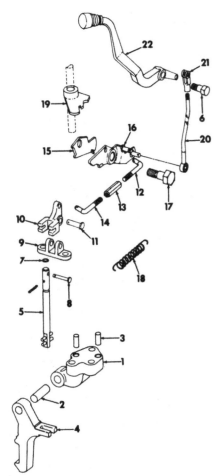

Fig. IH276—Exploded view of typical park lock and linkage. Park lock pawl (4) engages teeth on reverse driven gear in range transmission.

1. Bracket	12. Link
2. Pin	13. Turnbuckle
3. Dowel	14. Link
4. Pawl	15. Neutral pawl
5. Actuating shaft	16. Bellcrank
6. Bolt	17. Pivot bolt
7. "O" ring	18. Spring
8. Pin	19. Neutral plate
9. Bracket	20. Operating rod
10. Actuating lever	21. Clevis
11. Pin	22. Operating lever

Fig. IH278 — Side view showing correct installation of pinion shaft locating tool.

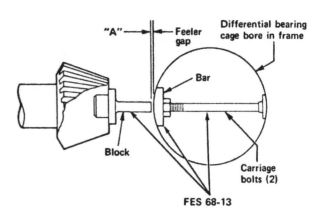

is contacting transmission cover, then while rocking rear wheel slightly, rotate turnbuckle (P) in direction of arrow until finger tight. Tighten jam nuts (N & O) and connect the actuating spring.

When park lock is properly adjusted and is in engaged position, a dimension of approximately $\frac{11}{16}$-inch should exist between range transmission cover and pin (Q) in actuating shaft. Reinstall support covers and TA control lever.

Series 756-2756-856-2856-1256-21256-1456-21456

243. To adjust the range transmission and park lock linkage, first remove the TA control lever and the steering support top and rear covers. Move range transmission control lever to neutral position. Refer to Fig. IH-281 and disconnect the Hi-Lo and the reverse ball joint linkages at studs (A). Remove transmission lower linkage pins (B). With park lock operating lever in disengaged (up) position, remove pin (G) and loosen jam nut (H). Hold park lock lever approximately ¾-inch from support housing as shown and with bellcrank (F) in disengaged stop position (against support base), adjust clevis (AB) until pin (G) can be installed. Secure pin with cotter pin and tighten nut (H).

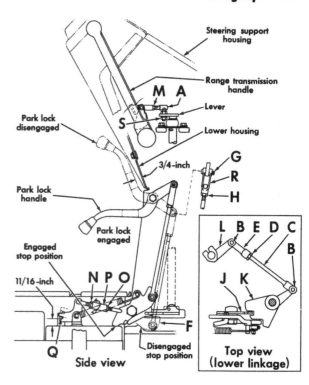

Fig. IH280 — View showing range transmission and park lock linkage adjustment on series 706, 2706, 806, 2806, 1206 and 21206. Refer to text for procedure.

Move park lock lever to engaged (down) position so that pawl (J) passes through the neutral notches in plates (K and S). With lever (L) in neutral position and pawl (J) centered in neutral notch in plate (K), adjust clevis (E) on reverse linkage (R) until pin (B) can easily be installed through clevis and lever (L). Secure pin and tighten jam nut (D). Then, with lever (T) in neutral position and pawl (J) centered in neutral

notch in plate (S) adjust clevis (E) on Hi-Lo linkage (C) until pin (B) can be installed through clevis and lever (T). Secure pin and tighten jam nut (D).

With range transmission control lever centered in neutral gate of shift pattern cover (W), align slots in Hi-Lo shift hub (AD) and reverse shift hub (Y) with shift pin (Z) on shift lever. Loosen jam nuts (M) and adjust both ball joint linkages so that

Early Setting Numbers	Late Setting Numbers	Feeler Gap "A"
	60	.006
	61	.009
	62	.012
	63	.015
	64	.018
	65	.021
	66	.024
4	68	.030
5	69	.033
6	70	.036
7	71	.039
8	72	.042
9	73	.045
10	74	.048
11	75	.051
12	76	.054
13	77	.057
14	78	.060
15	79	.063
16	80	.066
17	81	.069
18	82	.072
19	83	.075
20	84	.078
21	85	.081
22	86	.084
	87	.087
	88	.090
	89	.093
	90	.096

Fig. IH279 — Chart used in determining feeler gap "A". Setting numbers (4 thru 22 or 60 thru 90) are etched on pinion shaft.

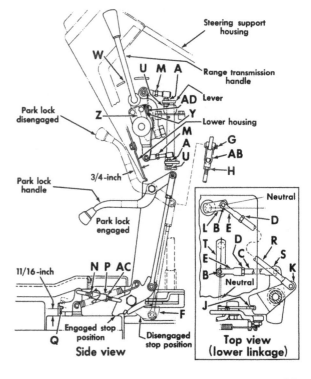

Fig. IH281 — View showing range transmission and park lock linkage adjustment on series 756, 2756, 856, 2856, 1256, 21256, 1456 and 21456. Refer to text for procedure.

studs (A) align with holes in levers on vertical shafts. Secure with nuts (U) and tighten jam nuts (M).

With park lock operating lever in engaged position, disconnect the actuating spring. Back off jam nuts (N and AC). Make certain that bellcrank

(F) is contacting transmission cover (engaged stop position), then while rocking rear wheel slightly, rotate turnbuckle (P) in direction of arrow until finger tight. Tighten jam nuts (N and AC) and connect the actuating spring.

When park lock is properly adjusted and is in engaged position, a dimension of approximately $1\frac{1}{8}$-inch should exist between range transmission cover and pin (Q) in actuating shaft. Reinstall steering support covers and TA control lever.

MAIN DRIVE BEVEL GEARS AND DIFFERENTIAL

The differential is carried on tapered roller bearings. The bearing on the bevel gear side of the differential is larger than that on the opposite side and therefore, it is necessary to keep the bearing cages in the proper relationship.

ADJUSTMENT

All Models

244. **CARRIER BEARING PRE-LOAD.** The carrier bearings can be adjusted by either of two methods; however, in either case the hydraulic lift assembly should be removed as outlined in paragraph 281, and the final drives removed as outlined in paragraph 250.

NOTE: After adjusting carrier bearings on late production series 756, 2756, 856 and 2856 or on all series 1206, 21206, 1256, 21256, 1456 and 21456, readjust bull pinion bearings as outlined in paragraph 253.

245. To use the direct measurement method, proceed as follows: Install differential in rear frame with no shims behind bearing cages and tighten left bearing cage cap screws to 75 ft.-lbs. torque. Rotate the differential and tighten the right hand bearing cage cap screws to 30-50 ft.-lbs. torque. Now loosen the right hand bearing cage cap screws, rotate differential and retighten cap screws to 20 in.-lbs., then without rotating differential, tighten bearing cage cap screws to 50 in.-lbs. torque. Use a depth gage through puller bolt holes and measure between the surface of rear main frame and outside of bearing cage and average the readings. Measure the thickness of bearing cage flange at puller bolt holes and average these readings. Subtract second reading from first reading which will give the required thickness of shim pack. Shim pack must be within plus or minus 0.002 of the determined shim pack thickness. Shims are available in thicknesses of 0.003, 0.004, 0.007, 0.012 and 0.030. Shims can be divided between the two bearing cages to pro-

vide the proper backlash as outlined in paragraph 247.

246. To use the rolling torque method, use the original shim packs behind bearing cages and rotate the differential as bearing cage cap screws are tightened. Be sure there is some backlash maintained between bevel gear and pinion. Place the range transmission in neutral, then loosen the right bearing cage so there is no preload on bearings (cap screws finger tight). Wrap a cord around differential, attach to a spring scale and note the pounds pull required to keep differential and transmission in motion. Tighten the bearing cage cap screws securely and recheck the rolling torque. Now vary the shims until 2 to 7 pounds more pull is required to keep differential and transmission in mo-

tion than when no preload was applied to bearings. With bearing preload determined, refer to paragraph 247 to set backlash between bevel gear and pinion.

247. **BACKLASH ADJUSTMENT.** With the differential carrier bearing preload determined as outlined in paragraph 245 or 246, the backlash between bevel gear and drive pinion should be checked and adjusted as follows: Mount a dial indicator and while holding drive pinion forward, check backlash in at least three places during a revolution of the differential. Correct backlash is 0.005-0.012. If the backlash is not as stated, shift bearing cage shims from one side to the other as required. NOTE: Do not add or remove shims as the previously determined bearing preload will be

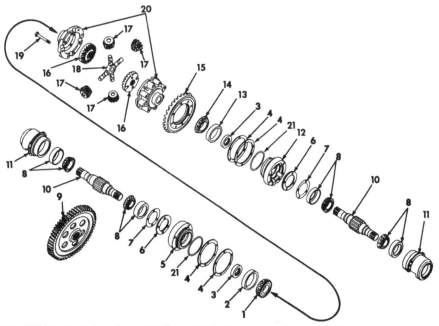

Fig. IH282—Exploded view of differential, bull pinions and associated parts used on early production series 756, 2756, 856 and 2856 and all series 706, 2706, 806 and 2806.

1. Bearing cone (LH)	7. Bull pinion bearing shim	13. Bearing cup (RH)
2. Bearing cup (LH)	8. Bearing	14. Bearing cone (RH)
3. Oil seal	9. Bull gear	15. Bevel ring gear
4. Bearing cage shims	10. Bull pinion shaft	16. Bevel gears
5. Differential bearing cage (LH)	11. Bull pinion bearing cage	17. Pinion gears
6. Retainer	12. Differential bearing cage (RH)	18. Spider
		19. Case bolt (8 used)
		20. Differential case
		21. "O" ring

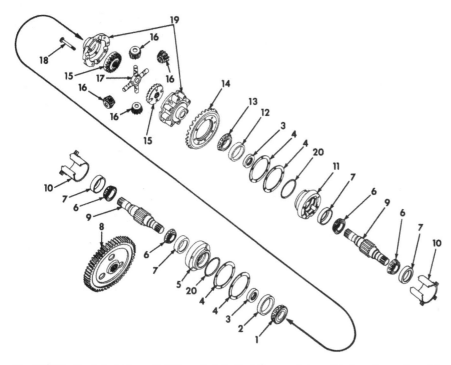

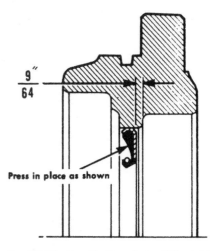

Fig. IH285—Install oil seals in differential bearing cages as shown on late production series 756, 2756, 856 and 2856 and all series 1206, 21206, 1256, 21256, 1456 and 21456.

Fig. IH283—Exploded view of differential, bull pinions and associated parts used on late production series 756, 2756, 856 and 2856 and all series 1206, 21206, 1256, 21256, 1456 and 21456.

1. Bearing cone (LH)	6. Bearing cone	13. Bearing cone (RH)
2. Bearing cup (LH)	7. Bearing cup	14. Bevel ring gear
3. Oil seal	8. Bull gear	15. Bevel gears
4. Bearing cage shims	9. Bull pinion shaft	16. Pinion gears
	10. Oil shield	17. Spider
5. Differential bearing cage (LH)	11. Differential bearing cage (RH)	18. Case bolt (8 used)
		19. Differential case
	12. Bearing cup (RH)	20. "O" ring

changed. Shifting 0.010 shim thickness from one side to the other will change backlash approximately 0.0075.

R & R BEVEL GEARS

All Models

248. The main drive bevel pinion is also the range transmission mainshaft. The procedure for removing, reinstalling and adjusting pinion setting is outlined in the range transmission section (paragraphs 237 thru 241).

To remove the bevel ring gear, follow the procedure outlined in paragraph 249 for R & R of differential. The ring gear is secured by the differential case bolts which should be tightened to a torque of 112 ft.-lbs.

R&R DIFFERENTIAL

All tractors are now equipped with a four pinion type differential. However, some series 706 and 2706 tractors were equipped with a two pinion type. Any difference in service procedure will be obvious during disassembly.

All Models

249. To remove differential, remove final drives as outlined in paragraph

250 and the hydraulic lift assembly as outlined in paragraph 281.

With final drives and hydraulic lift assembly removed, attach a hoist to differential assembly, remove both differential bearing cages and carefully lift differential from rear frame.

To disassemble differential, use a puller and remove the carrier bearings and note that bearing on bevel gear side of differential has a larger O. D. and contains more rollers than the opposite side bearing. Remove the differential case bolts and separate differential assembly. Any further disassembly is obvious. Oil seals and

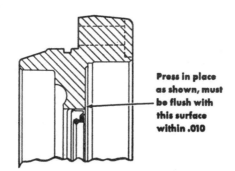

Fig. IH284—Install oil seals in differential bearing cages as shown on early production series 756, 2756, 856 and 2856 and all series 706, 2706, 806 and 2806.

bearing cups in bearing cages can also be renewed at this time and procedure for doing so is obvious. Oil seals are installed with lip toward inside and within 0.010 on early production series 756, 2756, 856 and 2856 and all series 706, 806, 2706 and 2806 or 9/64-inch on late production series 756, 2756, 856 and 2856 and all series 1206, 21206, 1256, 21256, 1456 and 21456, of being flush with outside surface of bore. See Fig. IH284 and IH285.

When reassembling differential, tighten the differential case bolt nuts to 112 ft.lbs. Refer to paragraphs 244 thru 247 for carrier bearing preload and backlash adjustments.

FINAL DRIVE

The final drive assemblies consist of the rear axle, bull gear and bull pinion and can be removed from the tractor as a unit.

All Models Except High Clearance

250. **REMOVE AND REINSTALL.** With the exception of the brake valve assembly, removal of either final drive assembly is the same. Therefore, removal procedure will be given for the right final drive assembly and will include the necessary brake valve removal procedure.

251. To remove the right final drive, first remove drain plug and drain housing. Remove fender, then support tractor under main frame and remove the tire and wheel assembly. To keep from bending lines, remove plat-

Fig. IH286 — Right final drive assembly removed from rear frame of Farmall 806. Other series are similar. Items (6) and (12) are not used on late production series 756, 2756, 856 and 2856 or series 1206, 21206, 1256, 21256, 1456 and 21456.

2. Bearing cone
3. Rear axle shaft
6. Bearing retainer
10. Bull gear
11. Bull pinion
12. Bearing cage

bearings in bearing cage, install inner bearing cup, then install retainer with one shim (0.007). Rotate bull pinion shaft while tightening retainer cap screws and after cap screws are tight, bump bull pinion shaft on both ends to insure that bearings are seated. Bull pinion shaft should rotate freely with no binding or excessive end play. If shaft binds, add one (0.007) shim and recheck. Press rear axle outer bearing on axle with largest diameter toward inside using a pipe having a diameter which will contact inner race only. Press bearing on axle shaft until it bottoms against shoulder. Install grease shield in carrier, then install axle. If oil seal in outer cap (retainer) is renewed, install with lip toward inside and inner surface flush with inner surface of bore as shown in Fig. IH289. Use shim stock or seal sleeve when installing cap over axle and tighten retaining cap screws finger tight only at this time. Install outer snap ring, bull gear, inner snap ring and inner bearing on inner end of axle. Install inner bearing cup in main frame if necessary. Use guide studs and install final drive assembly on rear main frame. Tighten three of the cap screws in axle outer cap to 150 in.-lbs., rotate axle to align and seat bearings, then loosen and retighten **evenly** to 50 in.-lbs. Measure distance between outer cap and carrier next to the three tightened cap screws and average the three readings. Make up a shim pack that will be within a plus or minus 0.002 of the average measured distance and install between axle outer cap and carrier. Tighten all cap screws securely and complete reassembly of tractor.

form, then disconnect and remove the brake valve lines. Unbolt brake valve from mounting bracket, swing valve forward, then unbolt mounting bracket from final drive housing and remove brake valve, bracket and brake pedals. Remove brake housing and brake assembly. Remove bolt circle from final drive housing, attach hoist and remove final drive assembly from main frame as shown in Fig. IH286.

Reinstall by reversing the removal procedure.

252. OVERHAUL (706-806-2706-2806 and early series 756-856-2756-2856). With final drive assembly removed as outlined in paragraph 251, disassemble as follows: Use a puller and remove bearing cone (2—Fig. IH-286) from inner end of axle (3) and remove snap ring. Remove cap screws from axle outer bearing retainer (16—Fig. IH288), then bump axle out of bull gear and remove bull gear. If axle is to be completely removed from the housing, it will be necessary to either remove the inner snap ring from axle or remove the grease shield

(7) in outer axle housing as the snap ring will not pass through the grease shield. Remove bearing retainer and shims and pull the bull pinion shaft and bearing from bearing cage.

NOTE: Unless outer bearing cup in bearing cage is to be renewed, or the bearing cage is damaged, there is no need to remove bearing cage from housing. Also note upper right view in Fig. IH288. Spacer (18) is used with a different carrier (21) when tractor is equipped with wide tread rear axles. Any difference in disassembly is obvious.

Any further disassembly required is self-evident after an examination of the unit and reference to Fig. IH288.

Reassemble final drive assembly as follows: Install outer bearing cup in bull pinion cage, if necessary, then install bearing cage in carrier. Install bearings on bull pinion shaft with largest diameters toward gear (inside). Place bull pinion shaft and

Fig. IH288 — Exploded view of final drive assembly. Note carrier (21) and extension (18) used when tractor has wide tread rear axles. Snap ring (17) is not used on series 1206, 21206, 1256, 21256, 1456 and 21456.

1. Bearing cup
2. Bearing cone
3. Snap ring
4. Gasket
5. Carrier
6. Rear axle shaft
7. Grease shield
8. Bearing cone
9. Bearing cup
10. Shim (light)
11. Shim (medium)
12. Shim (heavy)
13. Shim (ex. light)
14. Oil seal
15. Grease fitting
16. Cap
17. Snap ring
18. Extension
19. "O" ring
20. Plug
21. Carrier (wide tread)

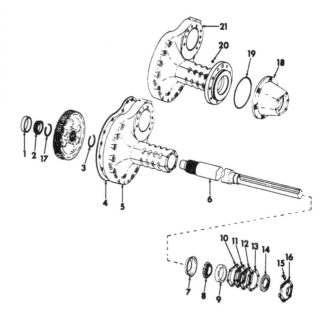

Fig. IH287—Right side view of rear main frame with final drive removed.

1. Axle inner bearing cup
13. Differential bearing cage
33. Cup plug

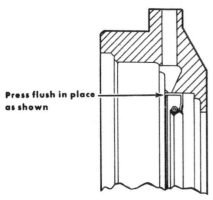

Press flush in place as shown

Fig. IH289 — When installing rear axle cap (retainer) seal, position as shown with lip toward inside.

253. OVERHAUL (1206-21206-1256-21256-1456-21456 & late series 756-2756-856-2856). With final drive removed as outlined in paragraph 251, disassemble as follows: Remove brake piston housing and adjusting shims, then remove the bull pinion. Using a suitable puller, remove bearing cone from inner end of axle shaft. Remove bull gear and bull gear retaining snap ring. Unbolt axle outer bearing retainer and withdraw axle shaft assembly from carrier. Any further disassembly required is evident after an examination of the unit.

When reassembling, press rear axle outer bearing on axle using a pipe having a diameter which will contact inner race only. Press bearing on axle shaft until it bottoms against shoulder. Install grease shield in carrier, if it was removed, then install axle. If oil seal in outer bearing retainer is renewed, install with spring loaded lip toward inside and inner surface flush with inner surface of bore as shown in Fig. IH289. Use shim stock or seal sleeve when installing seal retainer over axle. Install retainer cap screws but do not tighten them at this time. Install snap ring, bull gear and inner bearing on inner end of axle shaft. Place the bull pinion in position on bull gear and install brake piston housing without shims or "O" ring. Use aligning dowels and install the final drive assembly to the main frame. NOTE: Hold bull pinion in position while installing the assembly so seals and bearings will not be damaged.

On all series except 1456 and 21456, install three cap screws through brake piston housing and tighten them evenly to a torque of 100 in.-lbs. while rotating the bull pinion shaft to align and seat the bearings. Loosen the cap screws, then retighten them evenly to a torque of 50 in.-lbs. while rotat-

ing the shaft. Next, without rotating the bull pinion shaft, further tighten the cap screws to a torque of 100 in.-lbs. Using a feeler gage, measure the gap adjacent the cap screws. The correct shim pack is the average measured gap plus 0.015. Shims are available in the following thicknesses: 0.003, 0.004, 0.007, 0.012 and 0.035. Remove brake piston housing, install new "O" ring and oil seal, then reinstall brake piston housing with the shim pack.

On series 1456 and 21456, install and tighten two cap screws through brake housing, 180 degrees apart, evenly to a torque of 100 in.-lbs. while rotating the shaft. Loosen the two cap screws, then retighten them to 100 in.-lbs. while rotating the shaft. Using a feeler gage, measure the gap between brake housing and final drive housing, adjacent to the two cap screws. The correct shim pack is the average measured gap plus 0.009. Shims are available in the following thicknesses: 0.003, 0.004, 0.007, 0.012 and 0.035. Remove brake piston housing, install new "O" ring and oil seal, then rein-

stall brake piston housing using the correct shim pack.

To adjust the axle shaft bearings on all series, tighten three evenly spaced cap screws retaining the axle outer cap (seal retainer) to axle carrier to a torque of 150 in.-lbs. Rotate axle shaft to align and seat bearings, then loosen the cap screws and retighten evenly to 50 in.-lbs. Measure distance between outer cap and carrier next to the three tightened cap screws and average the three readings. Make up a shim pack that will be within a plus or minus 0.002 of the average measured distance and install between axle outer cap and carrier. Tighten all cap screws securely and complete reassembly of tractor.

High Clearance Models

254. On high clearance model tractors, the complete final drive assembly is removed in the same manner as outlined in paragraph 251 although the drop housing (18—Fig. IH290) should also be drained.

With unit removed, proceed as follows: Remove pan (19) from housing

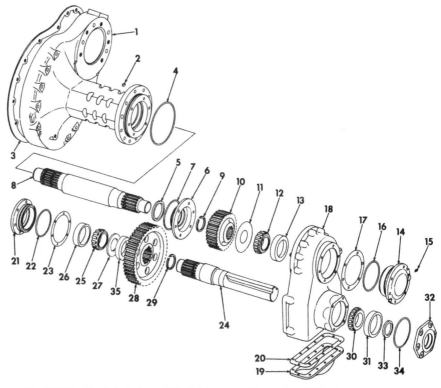

Fig. IH290—Exploded view of final drive assembly used on high clearance tractors.

1. Carrier	13. Bearing cup	25. Inner bearing cone
2. Plug	14. Bearing cap	26. Bearing cup
3. Gasket	15. Grease fitting	27. Axle spacer
4. "O" ring	16. "O" ring	28. Driven gear
5. Oil seal	17. Shim	29. Snap ring
6. Seal retainer	18. Housing	30. Outer bearing cone
7. "O" ring	19. Pan	31. Bearing cup
8. Drive gear axle	20. Gasket	32. Bearing cap
9. Snap ring	21. Bearing cap	33. Oil seal
10. Drive gear	22. "O" ring	34. "O" ring
11. Grease retainer	23. Shim	35. Gear spacer
12. Bearing cone	24. Rear axle shaft	

(18). Remove bearing cap assembly (items 21, 22, 23 and 26), then remove inner bearing (25) and spacers (27 and 35) from end of axle (24). Remove outer bearing cap assembly (items 32, 34, 33 and 31), then bump axle outward and remove driven gear (28) from bottom of housing. Remove bearing cap assembly (items 13, 14, 16 and 17) from housing (18), then unbolt and remove housing (18) from carrier (1). Remove bearing (12), grease retainer (11) and drive gear (10), then remove snap ring (9) and the oil seal retainer (6) and oil seal (5) assembly. Remove bearing and snap ring from inner end of axle and bump axle out of bull gear. Any further disassembly required will be obvious after

an examination of the unit and reference to Fig. IH290.

Clean and inspect all parts and renew as necessary. Be sure to use all new "O" rings and gaskets and reassemble by reversing the disassembly procedure. However, delay the final adjustment of shaft bearings until unit is installed on tractor. Refer to paragraph 255 for procedure.

255. To adjust the drive gear axle bearings, remove all the shims (17—Fig. IH290) from behind bearing cap (14). Install three cap screws (one every other hole) in bearing cap and tighten evenly to 150 in.-lbs. Bump bearing cap to insure bearings are seated, loosen the three cap screws, then retighten to 50 in.-lbs. Measure

distance between bearing cap and drop housing next to the three cap screws, then select and install a shim pack equal to the averaged reading. This will provide the shaft bearings with the recommended zero end play. Shims are available in 0.004, 0.005, 0.007, 0.012 and 0.0299 thicknesses.

Repeat the above operation to adjust bearings for the rear axle shaft (24).

NOTE: Seals (5) and (33) are installed with lips facing toward inside. Use shim stock or a seal protector when installing retainer (6) or cap (32).

Complete reassembly of tractor and bleed brakes as outlined in paragraph 258.

BRAKES AND CONTROL VALVE

Brakes on all models are actuated hydraulically and are a self-adjusting, double disc type. Brake operation can be accomplished with engine inoperative because of a one-way check valve located on top rear of the multiple control valve. This check valve closes when hydraulic pressure ceases and thus provides a closed circuit which permits operating the brake control valve with the oil trapped within the circuit.

Service (foot) brakes MUST NOT be used for parking or any other stationary job which requires the tractor to be held in position. Even a small amount of fluid seepage would result in brakes loosening and severe damage to equipment or injury to personnel could result. USE PARK LOCK when parking tractor.

BRAKE ADJUSTMENT

All Models

256. The only external adjustment that can be made on brakes are the brake pedal maximum travel and the brake control valve adjustment.

257. To make the brake pedal maximum travel adjustment, refer to Fig. IH291 and proceed as follows: Disconnect brake springs to simplify operation. Remove cotter pin (G) and pin (H), then move pedal forward until it is stopped by stop screw (F). At this time, rear of pedal arm should be 4⅛ inches on series 706, 2706, 806, 2806, 1206 and 21206 or 3½ inches on series 756, 2756, 856, 2856, 1256, 21256, 1456 and 21456, forward of the platform stop (C) as shown. If measurement (T) is not as stated, loosen jam nut and turn stop screw as required. Tighten jam nut and reinstall pin (H) and cotter pin (G). Repeat operation on other pedal. Do not reconnect re-

turn springs until brake valve is adjusted as follows:

To adjust brake valve, refer to Fig. IH292 and loosen cap screw (A). Use a small rod in hole of valve spool (B) and turn spool out of clevis until about ⅛-inch clearance appears between brake pedal arm and platform stop (C). Now, turn spool into clevis until pedal arm just contacts the pedal stop, then give valve spool an additional ¼-turn in the same direction. Tighten clamp screw (A). Repeat operation for other pedal and reconnect brake return springs.

BLEED BRAKES

All Models

258. To bleed brakes, attach a length of hose to bleeder valve (D—Fig. IH292) on top of brake housing and place open end in a container. Start engine and run at low idle rpm. Depress brake pedal and while holding

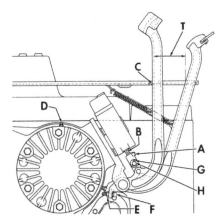

Fig. IH291 — View showing typical brake pedals, control valve and adjusting points. Distance (T) is measured with clevis pin (H) removed.

A. Clamp screw
B. Valve spool
C. Pedal return stop
D. Bleed valve
E. Jam nut
F. Pedal stop screw
G. Cotter pin
H. Clevis pin

Fig. IH292—View showing typical installation of brake valve on tractor.

A. Clamp screw
B. Valve spools
D. Bleed valve
H. Clevis pin
29. Pressure line
32. Left brake line
34. Right brake line

Fig. IH293—Brake housing used on all series except 1456 and 21456, with internal parts removed. Self-adjusting screw (14) has been removed for illustrative purposes.

10. Piston return springs
13. Primary plate
14. Self-adjusting screw
16. Intermediate plate
17. Brake disc

Fig. IH296—View showing proper installation of self-adjusting screw lock spring used on all series except 1456 and 21456.

in this position, open bleeder valve and when a solid flow of oil appears, close valve. Repeat operation on opposite brake. Check brake pedal feel. If brake pedal operation feels spongy rather than having a solid feel, repeat the bleeding operation.

NOTE: With engine not running, braking action should occur about midway of the pedal travel.

BRAKE ASSEMBLIES

All Models Except 1456-21456

259. **R & R AND OVERHAUL.** Removal of either brake is accomplished by removing cap screw and spacer located between platform and top of brake housing, disconnecting the brake line, then removing the housing retaining cap screws. Brake plates, return springs and discs can be lifted from housing. See Fig. IH293. Note

location of the self-adjusting screws as primary brake plate is removed. Piston housing and piston assembly can also be removed from final drive assembly.

Piston can be removed from piston housing by bumping edge of housing on work bench, or by applying air pressure to the line fitting of housing. However, when using air pressure to remove piston, use caution not to "blow" piston out which could result in damage to parts or injury to personnel. Should piston become cocked during removal, it will be necessary

to straighten it to complete removal. See Fig. IH294.

With brake assembly disassembled, clean and inspect all parts including cylinder cavity in piston housing. Install new "O" ring seals in piston housing and press piston into piston housing with insulator pad outward. Brake lining is bonded to discs and if renewal is required, renew complete disc. If oil seal is renewed in piston housing, install seal with lip toward inside and to dimension shown in Fig. IH295. When installing self-adjusting screw lock spring, refer to Fig. IH296. Be sure lock spring tab is in its hole and tighten spring retaining screw to 30 in.-lbs. torque. Coat threads of self-adjusting screw with Molykote EP or equivalent and while holding spring lock away from head of screw, turn screw in until head contacts primary plate and release lock. Note: After brake is installed and operated, adjusting screw will rotate and properly adjust brakes. Renew any of the brake (piston) return springs which are rusted, fractured or have taken a set and appear shorter than others.

Reassemble brakes by reversing disassembly procedure and be sure heads of the self-adjusting screws are toward the piston housing. See Fig. IH297.

When assembly is complete, start engine and bleed brakes as outlined in paragraph 258.

Fig. IH294 — Brake piston removed from piston housing. Note insulator (11) on piston.

3. Oil seal
4. Outer seal ring
5. Inner seal ring
11. Insulator
12. Piston

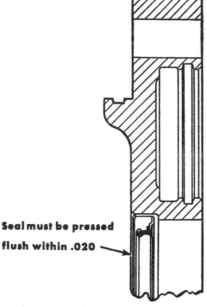

Seal must be pressed flush within .020

Fig. IH295—When installing oil seal in piston housing, install with lip inward (away from brake) and to dimension shown.

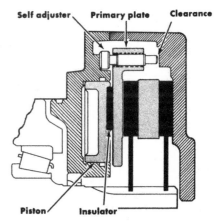

Fig. IH297—Cross-sectional view of brake assembly used on all series except 1456 and 21456.

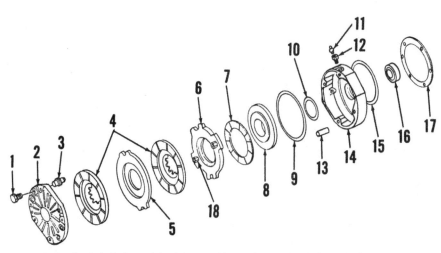

Fig. IH298—Exploded view of brake assembly used on series 1456 and 21456. Four self-adjusters (3) are used on each brake.

1. Adjuster carrier
2. Cover
3. Self-adjuster assy.
4. Brake disc
5. Intermediate plate
6. Primary plate
7. Insulator
8. Piston
9. Outer seal ring
10. Inner seal ring
11. Bleed valve
12. Connector
13. Dowel pin
14. Brake housing
15. "O" ring
16. Oil seal
17. Bull pinion bearing adjusting shims
18. Brake lining max. wear stop

Series 1456-21456

260. R&R AND OVERHAUL. To remove either brake, unbolt and remove cover (2—Fig. IH298). Adjuster carriers (1) and the four self adjuster assemblies (3) will be removed with cover. Brake discs (4), intermediate plate (5) and primary plate (6) can now be removed. Disconnect brake line, unbolt brake housing (14) and carefully withdraw the brake housing assembly. Do not lose shims (17) as these shims control bull pinion bearing adjustment.

Piston (8) can be removed from brake housing by bumping brake housing on work bench, or by applying air presure to the brake line fitting of housing. If air pressure is used, use caution not to "blow" piston out which could result in damage to parts or injury to personnel.

Clean and inspect all parts and renew any showing excessive wear or other damage. Install new oil seal (16) with lip toward inside (away from brake), then install new piston seal rings (9 and 10). Install piston (8) with insulator (7) in brake housing. Install brake housing using new "O" ring (15) and shim pack (17). Connect brake line.

Before installing the balance of brake parts, check and reset self-adjuster assemblies (3) as follows: Refer to Fig. IH299 and measure the distance plunger (P) protrudes from end of adjuster tube (T). This distance should be 0.025 to provide correct brake disc clearance when brakes are released. Reset the three snap rings

(R) so that inner snap ring is ¾-inch from end of adjuster tube (T) as shown. As brake linings wear, the primary plate will force adjuster tubes through snap rings and deeper into adjuster carriers (C) to maintain brake adjustment.

With self-adjusters reset and installed in cover as shown in Fig. IH-300, stack the balance of brake parts on cover as follows: Place outer brake disc in recess in cover, install intermediate plate, and inner brake disc, then install primary plate with lining maximum wear stops toward cover. Make certain that bosses on intermediate plate and primary plate straddle the dowel pins as shown. Install the assembly in brake housing.

When assembly is completed, check external adjustments as outlined in paragraph 256, start engine and bleed brakes as in paragraph 258.

Fig. IH300—View showing assembly of self-adjusters, brake discs, intermediate plate and primary plate ready for installation in series 1456 or 21456 brake housing.

Fig. IH299—Cross-sectional view of brake self-adjuster assembly used on series 1456 and 21456. Adjuster carriers (C) are screwed in from outside of brake cover.

C. Adjuster carrier
D. Pin
P. Plunger
R. Snap rings
S. Return spring
T. Adjuster tube

BRAKE CONTROL VALVE

All Models

261. R & R AND OVERHAUL. Remove brake control valve as follows: Disconnect all brake lines which in some cases will require removal of platform before dump line can be disconnected from valve body cap. Disconnect brake pedal return springs. Remove clevis pin from outer pedal clevis, then turn clevis one quarter turn to give clearance for removal of inner pedal clevis pin. Unbolt and remove control valve from mounting bracket.

262. With brake control valve removed, refer to Figs. IH301 and IH302 and disassemble valve as follows: Loosen the clevis clamp screws, hold valve spools with a small rod and count turns as each clevis is removed. This will eliminate a considerable amount of adjusting after control valve is reinstalled. Remove cap from valve body, then identify each spool with its bore and remove spool assemblies and springs. Place small rod,

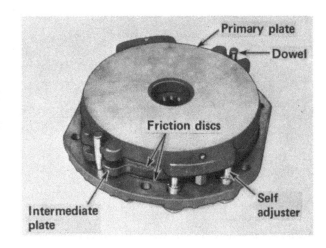

or pin punch, through hole in valve spool and remove the Phillips head screw, retainer (7) and reaction spring (3). The square cut snap ring (9) can also be removed if necessary. Remove the orifice (10).

Clean and inspect all parts. Pay particular attention to pistons (5 and 6) and their bores. Be sure orifice (10) is clean as well as all other oil passages. Renew springs (1 and 2) if any doubt exists as to their condition. Use all new "O" rings and gaskets and reassemble by reversing disassembly procedure.

NOTE: If reaction spring (3) and spring seat (7) were removed, a new Phillips head screw (PS—Fig. IH302) should be used during assembly. The Phillips head screw is sealed with "Loctite" when installed and a new screw and sealant is available as a service package under IH part number 384 608 R91.

After valve is installed, bleed brakes as outlined in paragraph 258 and adjust brake pedals as outlined in paragraph 257.

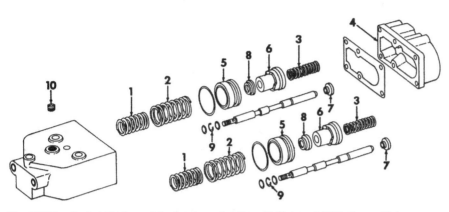

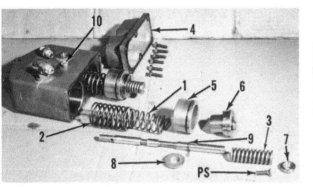

Fig. IH301—Exploded view of brake control valve. Gasket and "O" rings which are not numbered are available only in a kit, IH part No. 384502R92. Valve body and spools are not serviced separately.

1. Return spring
2. Large piston spring
3. Reaction spring
4. Body cap
5. Large piston
6. Small piston
7. Spool end
8. Spring seat
9. Snap ring
10. Orifice

Fig. IH302 — View showing brake control valve partially disassembled. When reassembling, use new Phillips head screws (PS) and coat threads with "Loctite". Refer to Fig. IH301 for legend.

BELT PULLEY

Belt pulley attachment is available for series 706, 2706, 756, 2756, 806, 2806, 856 and 2856. The belt pulley housing is attached to the PTO housing and the unit is driven by the 1000 rpm PTO output shaft.

REMOVE AND REINSTALL

All Models So Equipped

263. To remove the belt pulley assembly, disconnect inlet (top) hose from the orifice connector located at hole in PTO valve linkage support, remove orifice connector, then plug hole with 3/8-inch plug (with "O" ring). Disconnect outlet (bottom) hose from upper left side of PTO housing and plug hole with the 1/2-inch plug and "O" ring which is stored in a retainer located below the hole. Connect the disconnected ends of the two hoses together and tighten securely to retain oil in belt pulley housing. Remove two diagonally opposite retaining cap screws and install guide studs. Attach hoist to unit, remove the two remaining cap screws and slide unit rearward off the PTO output shaft.

Reinstall by reversing the removal procedure.

OVERHAUL

All Models So Equipped

264. With belt pulley removed, disassemble as follows: Refer to Fig. IH303. Remove retainer (1) and pull the shaft, gear and bearing assembly from housing. Bushing (12) and oil seal (13) can now be removed from housing. Remove pulley (26) from pulley shaft (25). Remove cotter key, nut and washer (Fig. IH304) from in-

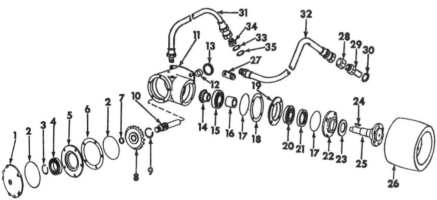

Fig. IH303—Exploded view of belt pulley attachment.

1. Bearing retainer
2. "O" ring
3. Snap ring
4. Roller bearing
5. Bearing cage
6. Shim
7. Plug
8. Drive gear
9. Snap ring
10. Input shaft
11. Housing
12. Bushing
13. Oil seal
14. Driven gear
15. Roller bearing
16. Spacer
17. "O" ring
18. Shim
19. Bearing cage
20. Roller bearing
21. Oil seal
22. Bearing retainer
23. Oil seal
24. Woodruff key
25. Pulley shaft
26. Pulley
27. Plug storage retainer
28. Flared cap
29. Orifice connector
30. "O" ring
31. Inlet hose
32. Outlet hose
33. "O" ring
34. Plug
35. "O" ring

ner end of pulley shaft (25—Fig. IH-303), then unbolt bearing retainer (22) and remove shaft and bearings assembly from housing. Any further disassembly required will be obvious.

Clean and inspect all parts. Pay particular attention to the straight roller bearings to see that they have no rough spots or are not excessively worn. Bushing (12) is pre-sized and if carefully installed should require no final sizing. Seals (13) and (23) are installed with lips toward inside of housing. Plug (7) in aft end of input shaft need not be removed unless damaged. Items (3, 4, 8 and 10) are available only as an assembly.

To reassemble, start with pulley shaft and reverse the disassembly procedure. Use all new "O" rings and seals. Vary shims (6) and (18) as required to provide 0.002-0.010 backlash between the gears. Shims are available in light, medium and heavy thicknesses.

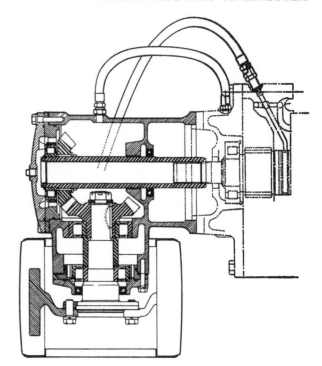

Fig. IH304 — Cross-sectional view of installed belt pulley attachment. Note lubrication hoses connected to PTO.

POWER TAKE-OFF

The power take-off used on all models is an independent type. On series 706, 2706, 756, 2756, 806, 2806, 856 and 2856, the pto is available as a dual speed unit having both 1000 and 540 rpm output shafts, or as a 1000 rpm unit which is convertible to the dual speed unit by adding the necessary 540 rpm shafts and gears. On series 1206, 21206, 1256, 21256, 1456 and 21456, the single speed 1000 rpm power take-off only is available. The pto unit for all models incorporates its own hydraulic pump which furnishes approximately 3 gpm of oil which is used to actuate the pto clutch and provide lubrication for the pto, and on tractors so equipped, the belt pulley unit. Operation of the pto unit is controlled by a spool type valve located in a bore in left side of pto housing.

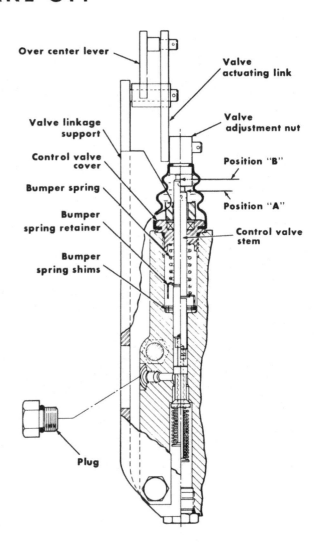

Fig. IH305 — Cut-away view of the PTO unit control valve. Note positions "A" and "B". Refer to text for adjustment.

OPERATING PRESSURE

All Models So Equipped

265. To check operating pressure, run tractor until hydraulic fluid is approximately 100 degrees F. Refer to Fig. IH305 and remove the test port plug which is located at hole in linkage support (left side) and attach either an IH Flow Rater, or a test gage capable of registering at least 300 psi. If necessary, loosen linkage support cap screws and push support downward as far as possible and retighten cap screws. Unscrew valve from adjusting nut and swing nut and actuating link up out of the way. Start

engine and with pto engaged, operate engine at 2100 rpm (1000 pto rpm). Pull valve stem up until stem contacts bumper spring retainer (position "A"—Fig. IH305) and hold in this pos-

ition. Note: This is partial engagement position. If gage reads 41-46 psi, pressure is satisfactory at this point. If pressure is below 41 psi, remove control valve cover, bumper spring

Fig. IH306 — PTO unit being removed. Note rear end being tipped downward to prevent damage to suction tube.

and bumper spring retainer and install bumper spring shims as required. Each shim will change pressure about 5 psi. Reposition the actuating link and adjusting nut and screw valve stem into nut to approximately the original position. Move control handle and pull the over-center link into up position (position B—Fig. IH305) and check pressure. Gage should read 175-185 psi on 706, 2706, 756, 2756, 806, 2806, 856 and 2856, 195-205 psi on 1206, 21206, 1256 and 21256 or 230-235 psi on 1456 and 21456. If gage pressure is not as stated, turn valve stem into adjusting nut to increase or out of nut to decrease pressure. Note: This is the full engagement position.

With these adjustments made, reduce engine speed to low idle rpm (pto engaged) at which time the pressure must not drop more than 40 psi

1. Extension shaft
2. Gasket
3. Safety shield
5. Protection tube
6. Suction tube
7. Seal
8. Housing
9. Plug
10. Steel ball
11. Steel ball
12. Plug
13. "O" ring
14. Plug
15. "O" ring
16. Bearing
17. Brake piston
18. "O" ring
19. "O" ring
20. Brake spring
22. Valve guide
23. Spring, light
24. Spring, heavy
25. Control valve
26. Guide stem
27. Dowel pin
28. Bumper spring
29. Shim
30. Sleeve spacer
31. Sleeve
32. Stop
33. Cover
34. Seal
35. "O" ring
36. Washer
37. Cover
38. Seal
39. Gasket
40. Plug
41. "O" ring
42. Linkage support
43. Over-center lever
44. Actuating link
45. Adjusting nut
46. Output shaft
　　(1000 rpm)
47. "O" ring
48. Snap ring
49. Bearing
50. Snap ring
51. Clutch cup
52. Seal (teflon)
53. Snap ring
　　(Truarc)
54. Piston
55. "O" ring
56. Seal
57. Driven disc
58. Drive disc
59. Retainer
60. Snap ring
61. Release spring
62. Spring retainer
63. Snap ring
64. Drive shaft
65. Bearing
66. Woodruff key
67. Carrier
68. Dowel
69. Bearing
70. "O" ring
71. Pump housing
72. Drive gear
73. Idler gear

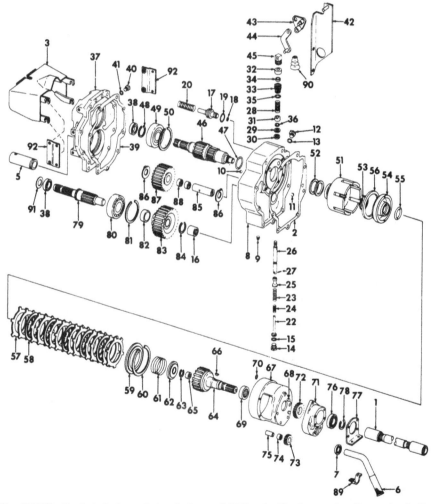

Fig. IH307—Exploded view of the dual speed PTO unit. Single speed units are basically similar except they do not include idler gear assembly (85 thru 88) or 540 rpm output shaft and gear unit (79 thru 84).

74. Bearing	80. Bearing	86. Thrust washer
75. Shaft	81. Snap ring	87. Idler gear
76. Bearing	82. Spacer	88. Bearing
77. Bearing retainer	83. Driven gear	89. Support
78. Snap ring	(540 rpm)	90. Dust boot
79. Output shaft	84. Snap ring	91. Seal shield
(540 rpm)	85. Idler shaft	92. Extension

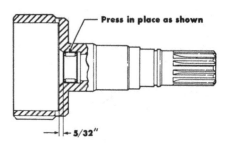

Fig. IH312—Needle bearing is pressed in bore of input shaft to dimension shown.

Fig. IH308 — Oil pump assembly which is mounted on front end of carrier assembly.

6. Suction tube	72. Drive gear
67. Carrier	73. Idler gear
71. Pump housing	76. Bearing

from the 175-185 psi reading on series 706, 2706, 756, 2756, 806, 2806, 856 and 2856, or the 195-205 psi reading on series 1206, 21206, 1256 and 21256 or the 230-235 psi reading on series 1456 and 21456. Disengage pto and check operation of pto brake (anti-creep). Brake must stop pto rotation within a maximum of 3 seconds.

If above conditions cannot be met, remove and overhaul pto unit as outlined in paragraph 266.

R&R AND OVERHAUL

All Models

266. To remove pto unit, drain the rear main frame, then disconnect control lever rod and remove linkage support. Attach hoist to unit, complete removal of retaining cap screws, then pull unit rearward, tip rear end downward to prevent damage to suction tube assembly and remove unit from tractor main frame. See Fig. IH306.

NOTE: If pto extension shaft remains with pto unit, it will be necessary to remove it from the pto unit input shaft before withdrawing unit from main frame.

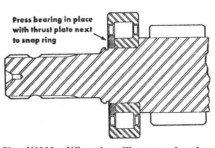

Fig. IH310—Rear side of housing showing brake pistons and springs.

17. Piston	19. "O" ring
18. "O" ring	20. Spring

Leave the two short cap screws which retain rear cover to housing in place so the pto sections will be held together.

267. With pto unit removed from tractor, use Fig. IH307 as a reference and disassemble unit as follows:

NOTE: This procedure is for the dual speed pto unit. Procedure for the single speed unit will remain basically the same except no idler gear and shaft, or 540 rpm driven gear and shaft, are used in the single speed unit.

Remove safety shield. Remove retainer (77—Fig. IH307), pump housing and suction tube assembly from carrier (67). See Fig. IH308. Remove pump idler gear from housing and the drive gear and Woodruff key from input shaft. If necessary, bearing (76) and suction tube (6) can be removed from pump housing. Note position of carrier (67) on housing (8—Fig. IH307), then unbolt and remove carrier. Remove the large snap ring (60), retainer (59) and the clutch discs (57 and 58). Place assembly in a press and using a straddle tool, depress clutch spring retainer (62), remove snap ring (63), then lift out retainer and clutch

release spring (61). Grasp hub of piston (54) and work piston out of clutch cup (51). Place a bar across clutch cup and again apply press to relieve pressure on the Truarc snap ring (53). Remove snap ring and pull clutch cup from the 1000 rpm shaft (46). Cover (37) can now be separated from housing (8) by removing the two short cap screws. See Fig. IH309. Shafts and gears are now available for service. Brake (anti-creep) springs and pistons can be removed from rear side of housing as shown in Fig. IH310. If control valve is to be disassembled, remove adjusting nut (45—Fig. IH-307), stop (32) and the valve cover (33) and seal (34) assembly. Remove plug (14) from bottom of housing and remove valve assembly. Pull spring (28), sleeve (31), washer (36), any shims (29) which may be present and spacer (30) from top of housing. Any further disassembly required will be obvious.

268. With unit disassembled, clean and inspect all parts and renew as necessary. Pay particular attention to

Fig. IH309—View after cover is removed showing arrangement of gears and shafts.

46. 1000 rpm shaft	85. Idler shaft
79. 540 rpm shaft	86. Thrust washer
83. 540 rpm gear	87. Idler gear

Fig. IH311—When installing rear bearing on 1000 rpm output shaft, thrust plate of bearing is next to snap ring.

Press bearing in place with thrust plate next to snap ring

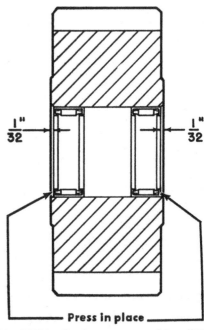

Press in place

Fig. IH313—Bearings are pressed into PTO idler gear 1/32-inch below flush.

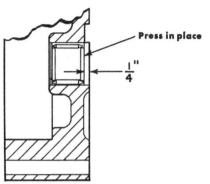

Fig. IH314—The 540 rpm shaft front bearing is pressed into housing to the dimension shown.

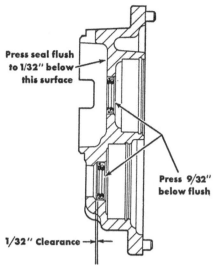

Fig. IH316—View showing seal installation in PTO cover. Lips are toward front.

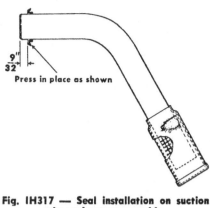

Fig. IH317 — Seal installation on suction tube and screen assembly.

the clutch discs which should be free of scoring or warpage. Use all new "O" rings, seals and gaskets during reassembly.

While reassembly is the reverse of disassembly, the following points are to be considered during reassembly.

When installing rear cover (37—Fig. IH307) to housing (8), the brake springs (20) will hold the sections apart so it will be necessary to use longer cap screws to pull the sections together. After sections are mated, the original cap screws can be installed. Be sure step on end of idler shaft (85) is toward front of pto unit as shaft mates with edge of carrier (67).

If the Teflon seal rings (52) used on hub of clutch cup are renewed, stretch the seals on the clutch cup. After the seals are installed, clamp a ring compressor around the seals to force them back to their original size.

When installing clutch piston (54) in clutch cup (51) be sure "O" ring is in inner bore of piston and the outer seal (56) has the lip facing away from piston hub. Use the following method to install piston in clutch cup. Obtain a piece of shim stock 4 inches wide, 18 inches long

and not more than 0.002 thick. Smooth all edges of the shim stock to preclude any possibility of injuring piston outer seal, then roll the shim stock and place it in clutch cup and against bottom of cup. Lubricate piston and inside surface of shim stock liberally with oil, then start piston into clutch cup and as piston is moved into position, carefully maintain the shim stock as nearly cylindrical as possible. Remove shim stock after piston is bottomed.

Clutch pack consists of 12 steel driven discs and 5 bronze drive discs. Install discs as follows: Start with two steel discs, then add one bronze disc. Repeat this 2 steel, 1 bronze assembly until the 12 steel and 5 bronze discs

are installed. When properly assembled, clutch pack will start and end with two steel discs. Install disc retainer (59) and snap ring (60). Use input shaft (64) to align discs.

When renewing bearings and seals, refer to Figs. IH311 through IH317 for dimensional information.

PTO DRIVEN GEAR

All Models

269. The pto driven gear, located at front of clutch housing, receives its drive from a hollow shaft which is splined to the backplate (cover) of the engine clutch. Removal of the pto driven gear requires that the tractor be split between rear frame and clutch housing; at which time, gear and bearing assembly can be removed from bottom of clutch housing.

After tractor has been rejoined, bearing of the pto driven gear should be checked, and if necessary, adjusted.

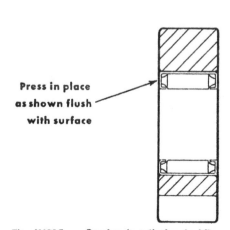

Fig. IH315 — Bearing installation in idler gear of PTO hydraulic pump.

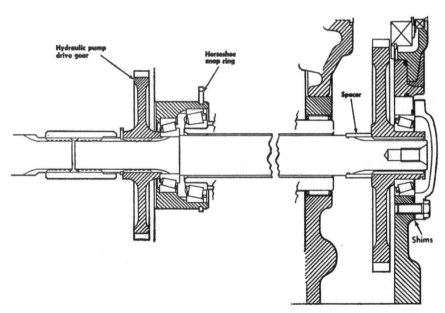

Fig. IH318—Schematic view showing PTO driven gear and shaft. Note location of adjusting shims.

Refer to Fig. IH318 for a schematic view showing arrangement of pto shafts. To adjust bearing, use only two cap screws and install bearing retainer without shims or seal. Turn gear and shaft and tighten the cap screws evenly to 10 in.-lbs. torque; then with out turning gear and shaft, tighten the cap screws to a torque of 20 in.-lbs. Now take three measurements around bearing retainer and average these readings. See Fig. IH319. Correct shim pack is gap (Z) plus 0.018-0.023. Shims are available in thicknesses of 0.007 and 0.028. Remove retainer, install shims and seal ring and tighten retainer cap screws.

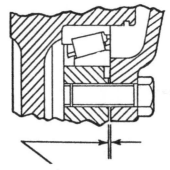

"Z" (Average of 3 readings)

Fig. IH319 — When adjusting bearing of PTO driven gear, measure gap (Z) as shown. Correct shim pack is gap (Z) plus 0.018-0.023.

HYDRAULIC LIFT SYSTEM

The hydraulic lift system provides load (draft) and position control in conjunction with either a 2-point or 3-point hitch on series 706, 2706, 756, 2756, 806, 2806, 856 and 2856 or the 3-point hitch only on series 1206, 21206, 1256, 21256, 1456 and 21456. Load control is taken from the lower links and transferred through a torsion bar and sensing linkage to the main control valve located in the hydraulic lift housing. Torsion bar and sensing linkage are located in the rear main frame. The hydraulic lift housing, which also serves as the cover for the differential portion of the tractor rear main frame, contains the work cylinder, rockshaft, valving and the necessary linkage. Mounted on the top right side of the lift housing is a seat support which contains the system relief valve and unloading valve along with control quadrant and levers. Also attached to the inside surface of the seat support, if tractor is equipped with an auxiliary system, are the auxiliary valves (either one or two) which control the hydraulic power to trailed or front mounted equipment. The pump which supplies the hydraulic system is attached to a plate which is mounted on left side of tractor rear main frame. The pump is driven by a gear located in the rear compartment of main frame on aft end of the pto driven shaft. All oil used by the hydraulic components of the tractor is drawn through a filter located in the rear main frame directly across from the hydraulic lift system pump.

TROUBLE SHOOTING

All Models

270. The following are symptoms which may occur during the operation of the hydraulic lift system. By using this information in conjunction with the Check and Adjust information and the R&R And Overhaul information, no trouble should be encountered in servicing the hydraulic lift system.

1. Hitch will not lift. Could be caused by:
   a. Unloading valve orifice plugged or piston sticking.
   b. Unloading valve ball not seating or seat loose.
   c. Faulty main relief valve.
   d. Faulty cushion relief valve.
   e. Faulty or disconnected internal linkage.
2. Hitch lifts when auxiliary valve is actuated. Could be caused by:
   a. Unloading valve orifice plugged.
   b. Unloading valve piston sticking.
   c. Unloading valve ball not seating or valve seat loose.
3. Hitch lifts load too slowly. Could be caused by:
   a. Unloading valve seat leaking.
   b. Excessive load.
   c. Faulty main relief valve.
   d. Faulty cushion relief valve.
   e. Scored work cylinder or piston or piston "O" ring faulty.
4. Hitch will not lower. Could be caused by:
   a. Drop piston sticking or "O" ring damaged.
   b. Control valve spool sticking or spring faulty.
   c. Drop check valve piston sticking.
5. Hitch lowers too slowly. Could be caused by:
   a. Action control valve spool sticking.
   b. Action control valve linkage maladjusted.
   c. Drop check valve "O" ring damaged or pilot ball cage maladjusted.
6. Hitch lowers too fast with position control. Could be caused by:
   a. Action control valve malfunctioning.

7. Hitch will not maintain position. Could be caused by:
   a. Drop check valve in main control valve leaking.
   b. Work cylinder or piston scored or piston "O" ring damaged.
   c. Check valve pilot valve leaking or ball cage maladjusted.
   d. Cushion valve leaking or damaged.
8. Hydraulic system stays on high pressure. Could be caused by:
   a. Linkage maladjusted, broken or disconnected.
   b. Auxiliary valve not in neutral.
   c. Mechanical interference.
9. Hitch over-travels (depth variation). Could be caused by:
   a. Torsion bar linkage sticking and needs lubrication.
   b. Unloading valve orifice partially plugged.
10. Draft (load) sensing too rapid in slow action position. Could be caused by:
   a. Action control linkage improperly adjusted.

TEST AND ADJUST

All Models

Before proceeding with any testing or adjusting, be sure the hydraulic pump is operating satisfactorily, hydraulic fluid level is correct and filter is in good condition. All tests should be conducted with hydraulic fluid at operating temperature which is normally 120-180 degrees F. Cycle system if necessary to insure that system is completely free of air.

271. **RELIEF VALVE.** On tractors equipped with an auxiliary hydraulic system, the relief valve can be tested as follows: Attach an IH Flo-Rater, or similar flow rating equipment, to any convenient outlet from an auxiliary control valve and be sure outlet hose from test unit is securely fastened in the hydraulic system reservoir. Start engine and run at rated speed. Manually hold auxiliary control valve in operating position, close valve of test unit and note the gage reading which should be 1450-1700 psi for series 706, 2706, 756, 2756, 806, 2806, 856 and 2856 and 1900-2100 psi for series 1206, 21206, 1256, 21256, 1456 and 21456. If pressure is not as stated, renew relief valve which is available only as a unit.

At this time, the hydraulic lift system pump delivery and the auxiliary control valve detent (latching) mechanism can also be checked.

At the engine rated rpm on series 706, 2706, 756, 2756, 806, 2806, 856 and 2856, the hydraulic lift pump should

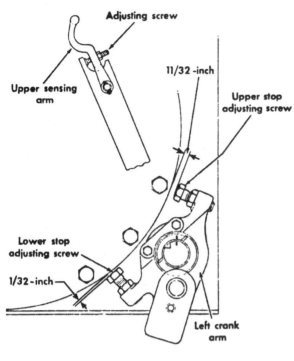

Fig. IH320 — Torsion bar left crank arm showing stop screw adjustments on series 706, 2706, 756, 2756, 806, 2806, 856 and 2856.

272. QUADRANT LEVERS. To test and adjust system, first check quadrant levers (load and position control). These levers should require 4-6 pounds of force, applied at knob, to move levers. If adjustment is necessary, adjust load control lever with nut on outer end of lever shaft. This nut is reached through a hole at bottom of quadrant. Adjust position control lever by working through opening in bottom of quadrant support. Loosen lock (ring) nut and turn the inner nut as required. A small punch can be used in the holes provided in nuts to make the position control lever adjustment.

273. CONTROL LINKAGE. To adjust the control linkage, first check and adjust, if necessary, the clearance for the torsion bar crank arm stop bolts. On series 706, 2706, 756, 2756, 806, 2806, 856 and 2856, refer to Fig. IH320 and adjust lower stop screw until the distance between head of stop screw and final drive housing is $\frac{1}{32}$-inch. Adjust upper stop screw to obtain a clearance of $\frac{11}{32}$-inch as shown.

On series 1206, 21206, 1256, 21256, 1456 and 21456 refer to Fig. IH321 and adjust the left stop screw until a clearance of $\frac{5}{16}$-inch exists between head of stop screw and left crank arm. Adjust right stop screw to obtain a clearance of $\frac{11}{32}$-inch between head of stop screw and right crank arm.

deliver 12 gpm at 1450-1700 psi.

On series 1206, 21206, 1256, 21256, 1456 and 21456 tractors, the hydraulic pump should deliver 12 gpm at 1900-2100 psi with engine operating at rated rpm.

To check the auxiliary control valve detent assembly, run engine at low idle rpm, pull control valve lever into operating position until it latches, then slowly close shut-off valve of the test unit and observe the pressure gage. Valve control lever should unlatch and return to neutral at not less than 1550 psi nor more than 1750 psi on series 1456 and 21456, or not less than 1000 psi nor more than 1250 psi

on all other series. If detent assembly does not operate properly, refer to paragraph 296.

NOTE: On tractors which are not equipped with auxiliary control valves, it will be necessary to remove the relief valve and bench test it in order to check its condition. To check the relief valve in this manner will require use of a hydraulic hand pump, gage and a test body such as an IH FES64-7-1 test body. Bear in mind that a relief valve tested in this manner will show a test pressure that will be on the low side of the pressure range due to the low volume of oil being pumped.

On all series, make certain there is no weight on hitch, then proceed as follows: Move position control lever forward to "LOWER" position (at offset in quadrant before action control section). Remove upper link bracket (3-point hitch) or plate (2-point hitch) to allow access to the sensing arm. Turn sensing arm adjusting screw in (clockwise) 3 or 4 turns. Place the load (draft) control lever $2\frac{7}{8}$ inches from rear of slot in quadrant on series 706, 2706, 756, 2756, 806, 2806, 856 and 2856 or 2 inches from rear of slot on all other series. Start engine and turn sensing arm adjusting screw out (counter-clockwise) until the hitch starts to raise. NOTE: Hitch will raise all the way. Install upper link bracket or plate.

Move load control lever to full forward position in slot on quadrant. Slowly move position control lever from rear of quadrant forward until hitch reaches fully lowered position. Remove seat support side cover and adjust turnbuckle (A—Fig. IH322) until position control lever is at the offset "LOWER" position on the quadrant. CAUTION: With seat support side cover removed, run engine at low idle rpm only. Install cover.

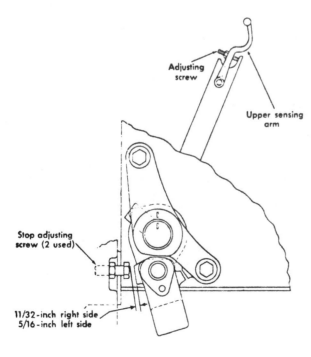

Fig. IH321 — Torsion bar crank arm stop screws are adjusted to a clearance of 11/32-inch on right side and 5/16-inch on left side on series 1206, 21206, 1256, 21256, 1456 and 21456.

Fig. IH322 — View showing location of turnbuckle (A) in right hand seat support.

With load control lever fully forward, move position control lever toward "RAISE" position until the following measurements between centerline of hole in outer right torsion bar crank arm and centerline of hole in rockshaft lift arm are obtained:
Series 706-2706-756-2756-806-2806-856-2856

    2-point hitch ....37⅞ inches
    3-point hitch ....39¾ inches
Series 1206-21206-1256-21256-1456-21456

    3-point hitch ....39$\frac{3}{16}$ inches
When correct measurement is obtained, adjust stop (S—Fig. IH323) against control lever and tighten securely.

Move position control lever to bottom of action control section (fully forward). Remove cover from front of lift housing. Adjust nut (A—Fig. IH324) until bushing is bottomed against valve body and continue turning nut (A) until position control lever moves ⅛-inch rearward away from front end of slot. Install the cover.

274. DROP CHECK VALVE SEAT ADJUSTMENT. The drop valve plug (30—Fig. IH325) is bottomed in the seat and must be left in that position.

Check and adjust the drop valve seat as follows: Attach an implement, or weights, to hitch and start engine. Place the load control lever in the heavy (forward) position. Move the position control lever toward raise until hitch has taken the attached weight, then move lever about 1-inch more in the same direction and mark the position of lever on quadrant. Now slowly move lever toward lower position until hitch just starts to lower. Measure distance the position control lever has moved. This distance should be ⅜-inch. If measurement is not ⅜-inch, remove front cover from lift housing and adjust drop valve seat as follows: Loosen lock nut and turn valve seat in (clockwise) if measurement was more than ⅜-inch; or turn seat out (counter-clockwise) if measurement was less than the ⅜-inch.

Turn seat in increments of ¼-turn and note that the drop valve seat has the two slots on its outer diameter.

275. POSITION CONTROL AND CYCLE TIME. With the above tests and adjustments made, hitch can be final checked for accuracy of position and the raise and lower times.

276. To check for position control accuracy, place a mark about midway of the quadrant, start engine and raise hitch by moving position control lever to rear of quadrant. Move position control lever forward to the affixed mark and measure distance from ends of lower links to ground. Repeat the operation and again measure the distance from ends of lower links to ground. These two measurements should not vary more than ⅛-inch. Now push position control lever forward to lower hitch, then move it rearward to the affixed mark. The measurement between ends of lower links and ground should not vary more than 1-inch from the measurements obtained in the first test. If differences are excessive, refer to paragraph 280.

277. To check hitch raise and lower times, be sure hydraulic fluid is at operating temperature and load hitch with an implement or weights. To check raise time, start engine and run at high idle rpm. Be sure hitch is in lowest position, then quickly move position control lever rearward to raise position. Hitch should reach full raise in three seconds or less.

278. To check the minimum lowering (drop) time, move the load control lever to the light load (rear) position and the position control lever forward to offset of quadrant. Now move the load control lever forward to the heavy position. Hitch should completely lower in two seconds or less.

279. To check the maximum lowering (drop) time, repeat the minimum time operation given in paragraph 278 except the position control lever should be in the extreme forward slow action position. Hitch drop time should be six seconds or more.

280. If position control accuracy and the hitch raise and lower times are not satisfactory, remove lift housing from rear main frame and inspect internal linkage. A visual inspection is sufficient to find any linkage defects. Renew any linkage or pins showing excessive wear or damage. Also check the main valve return spring and the main valve spool for binding. Spring can be renewed; however, if valve spool is defective, renew the spool and body assembly.

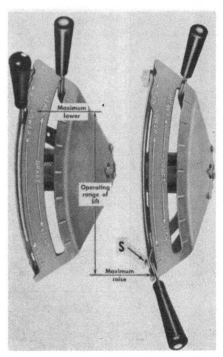

Fig. IH323 — When maximum raise measurements are correct, secure stop (S) against position control lever.

HYDRAULIC LIFT HOUSING

All Models

281. R&R AND OVERHAUL. To remove the lift housing, remove seat and platform, then if tractor is so equipped, disconnect the flexible hoses from the auxiliary valve lines. Disconnect lift links from rockshaft arms and remove rockshaft arms. Disconnect the rear break-away coupling bracket and the pto lever control rod. Remove attaching cap screws and attach a hoist to housing so that unit

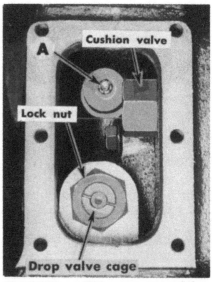

Fig. IH324 — With cover removed from front of lift housing, units shown are available for adjustment or service. (A) is esna nut on forward end of action control valve actuating rod.

will balance as it is removed. Carefully lift assembly from main frame and if available, mount unit in an engine stand.

NOTE: At this time, quadrant control levers and sensing pick-up arm (68—Fig. IH326) are susceptible to damage. If no engine stand is available, it is recommended that the right seat support, control lever quadrant assembly and the auxiliary control valves and tubing (if so equipped) be removed as a unit. The lower ends of control lever rods must be disconnected before assembly can be removed from lift housing. Also remove the left seat support.

282. Disconnect and remove sensing (return) spring (71), then remove pick-up arm (68). Disconnect link (56) from main control valve spool. The complete valve and work cylinder assembly can now be unbolted and removed from housing.

With assembly removed, the action control valve can be removed from main control valve and main control valve removed from the cylinder and the procedure for doing so is obvious.

When reinstalling cylinder and valve assembly in housing, install the mounting cap screws loosely; then use a small bar to hold cylinder assembly against mounting boss in housing and tighten the mounting cap screws securely. Complete reassembly by reversing the disassembly procedure and adjust system as outlined in paragraphs 271 through 279.

283. WORK CYLINDER AND PISTON. To disassemble the cylinder assembly after lift housing is removed as outlined in paragraph 281, first remove the lubrication tube, then remove piston (8—Fig. IH325) by bumping open end of cylinder against a wood block. Cushion valve (11) and the check valve assembly (items 4, 5, 6 and 7) can also be removed. If necessary, lever pin (3) can also be renewed.

Inspect cylinder and piston for scoring or wear. Small defects may be removed using crocus cloth. Pay particular attention to the piston "O" ring and back-up ring as well as the check valve poppet. If any doubt exists as to the condition of these items, be sure to renew them during assembly. Cushion valve (11) can be bench tested by using a hydraulic hand pump, gage, test body IH No. FES64 - 7 - 1, adapter IH No. FES 64-7-2 and petcock IH No. FES 64-7-4. Valve should test 1600-1800 psi on series 706, 2706, 756, 2756, 806, 2806, 856 and 2856 tractors or 2150-2350 psi

on series 1206, 21206, 1256, 21256, 1456 and 21456 tractors. Cushion valve adjusting screw is heavily staked in position and valve cannot be disassembled. If valve does not meet the above specifications renew the complete valve.

NOTE: When checking valve with hand pump bear in mind that the relief pressure obtained will be on the low side of the pressure range due to the low volume of oil being pumped.

Coat parts with IH "Hy-Tran" fluid before reassembly. Piston "O" ring is installed nearest closed end of piston.

284. MAIN CONTROL VALVE. To disassemble the main control valve after lift housing is removed as outlined in paragraph 281, first remove plug (30—Fig. IH325), spring (29) and ball (28) from end of check valve ball seat. Loosen lock nut (35), then while counting the number of turns, remove the valve seat. Remove snap ring (20) and pull spool assembly (16) from valve body. Compress return spring, remove snap ring (18) and disassemble spool assembly. Remove snap ring (25) and pull plug (24), spring (23) and poppet valve (21) from valve body.

Clean and inspect all parts. Spool return spring (17) has a free length of 2-21/64 inches and should test 18.4-21.6 pounds when compressed to a length of 1$\frac{29}{32}$ inches. Check ball spring

(29) has a free length of 59/64-inches and should test 3.5-4.1 pounds when compressed to a length of ¾-inch. Drop poppet valve spring (23) has a free length of 41/64-inch and should test 9-11 pounds when compressed to a length of 17/32-inch. If valve spool or spool bore show signs of scoring or excessive wear, renew complete valve assembly. All other parts are available for service.

Use all new "O" rings, dip all parts in IH "Hy-Tran" fluid and reassemble by reversing the disassembly procedure, keeping the following points in mind. Retainer (19) is placed on actuator (16) with relieved (chamfered) side toward spring (17). Drop valve seat is turned into valve body the same number of turns as were counted during removal. Tighten the lock nut (35) to 40-50 ft.-lbs. torque and plug (30) to 8-10 ft.-lbs. torque.

After assembly, mount valve on cylinder and piston assembly.

285. ACTION CONTROL VALVE. To disassemble the action control valve after lift housing is removed as outlined in paragraph 281, count the turns and remove the esna nut from end of actuating rod (46—Fig. IH325) and remove rod and spring (47). Remove retainer ring (45), then remove bushing (44), return spring (43) and variable orifice spool (42). The "O" ring (48) is located in I. D. of bushing bore. Remove plug (40), spring (39) and the drop control piston (38). Re-

Fig. IH325 — Exploded view of work cylinder, action control valve and main control valve assemblies.

2. Cylinder
3. Lever pin
4. Check valve poppet
5. Spring
6. Plug
7. "O" ring
8. Piston
9. "O" ring
10. Back-up washer
11. Cushion valve
12. "O" ring
14. "O" ring
15. "O" ring
16. Spool actuator
17. Spring
18. Retaining ring
19. Spring retainer
20. Snap ring
21. Drop poppet valve
22. "O" ring
23. Spring
24. Valve plug
25. Snap ring
26. "O" ring
27. Drop valve seat assy.
28. Ball
29. Spring
30. Plug
31. "O" ring
32. "O" ring
33. "O" ring
34. Actuating rod
35. Lock nut
37. "O" ring
38. Piston
39. Spring
40. Plug
41. "O" ring
42. Variable orifice spool

43. Return spring
44. Bushing
45. Snap ring
46. Actuating rod

47. Spring
48. "O" ring
49. Bellcrank
50. Snap ring

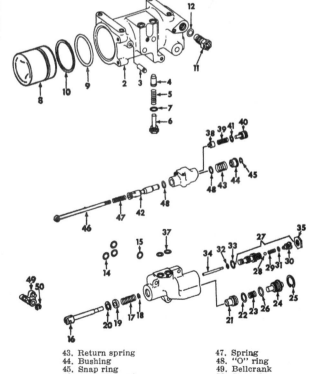

move "O" ring from I. D. of orifice spool bore.

Clean and inspect all parts and renew any which show signs of excessive scoring or wear. Spool (42) and piston (38) should be a snug fit yet slide freely in their bores.

Use all new "O" rings, dip all parts in IH "Hy-Tran" fluid and reassemble by reversing the disassembly procedure. Count the turns when installing the esna nut on end of actuating rod so nut will be installed to original position.

After assembly, mount action control valve on the main control valve.

286. ROCKSHAFT. Renewal of rockshaft or rockshaft bushings requires removal of the hydraulic lift housing; however, if only the rockshaft seals are to be renewed, it is possible to pry seals out of their bores

after removing the rockshaft lift arms. Seals are installed with lips toward inside and are driven into bores until bottomed. Dip seals in IH "Hy-Tran" fluid prior to installation.

With the lift housing removed, the rockshaft can be removed as follows: Remove set screws (76 and 82—Fig. IH326) and slide rockshaft out right side of housing. Bushings (2 and 3) can now be driven out of housing.

Rockshaft O. D. at bearing surface is larger on right side than on left. Rockshaft bearing surface O. D. is 3.020-3.022 for right side and 2.780-2.782 for left side. Rockshaft bushing I. D. is 3.023-3.028 for right side and 2.783-2.788 for left side. Rockshaft has 0.001-0.008 operating clearance in bushings. Use a piloted driver when installing bushings and install bushings with outer ends flush with bottom of seal counterbore.

When installing rockshaft, start rockshaft in right side of housing and align master splines of rockshaft, actuating hub (80) and bellcrank (75). Tighten hub and bellcrank set screws.

287. TORSION BAR ASSEMBLY. To remove the torsion bar assembly, from series 706, 2706, 756, 2756, 806, 2806, 856 and 2856 tractors, it is necessary to drain the rear main frame and remove the pto unit, if so equipped. Remove "U" bolt (19—Fig. IH328) and remove sensing arm assembly. Remove retainer (12) and left hand crank arm (9). Remove anchor bracket (6) and be sure not to lose retainer (8). Right hand crank (5) and torsion bar (7) assembly can now be withdrawn from rear frame. Do not let torsion bar slide out of right hand crank. Any chips, dents or scratches could set up stress points which might cause a torsion bar failure. Bushing case (1), bushing liner (2), "O" ring retainer (3) and "O" ring (4) renewal can now be accomplished and procedure for doing so is obvious.

Reassemble by reversing the disassembly procedure.

288. TORSION BAR ASSEMBLY. To remove the torsion bar assembly from series 1206, 21206, 1256, 21256, 1456 and 21456 tractors, first drain the rear main frame and remove the pto unit, if so equipped. Remove "U" bolt (16—Fig. IH329) and remove the draft sensing arm assembly. Unbolt and remove the torsion bar anchor bracket (7), then remove retainer (9) and right crank arm (5). Withdraw the left crank arm (10), crank arm shaft (6) and torsion bar (8) assembly from left side of main frame. Remove retainer (11) and separate the torsion bar from crank arm shaft and crank arm shaft from left crank arm. NOTE:

Fig. IH326—Exploded view of hydraulic lift housing showing relative position of component parts.

1. Lift housing	21. Position control lever
2. Bushing, LH	22. Knob
3. Bushing, RH	23. Guide
4. Oil seal, LH	24. Stop pin
5. Oil seal, RH	25. Knob
6. Bushing	26. Stop
7. Gasket	27. Friction disc
8. Cover	28. Position control shaft
9. Gasket	29. Spring
10. Seat support, RH	30. Nut
11. "O" ring	31. Lever
12. "O" ring	33. Load control lever
13. Quadrant support	34. Knob
14. Bushing	35. Friction disc
15. Oil seal	36. Load control shaft
16. Gasket	
17. Cover	37. "O" ring
18. Gasket	39. Lever
19. Quadrant	41. Support
20. Bearing	42. Lever
	43. Snap ring
	44. Bellcrank
	46. Snap ring
	47. Load control eccentric
	48. Retaining ring
	50. Position control eccentric
	51. "O" ring
	54. Link
	56. Link
	57. Pin
	59. Shoulder screw
	60. Walking beam
	62. Link
	64. Lower rod
	65. Turnbuckle
	66. Upper rod
	68. Sensing pick-up arm
	69. Pin
	71. Return spring
	72. Spring anchor
	73. Link
	74. Rockshaft
	75. Bellcrank
	76. Set screw
	77. Connecting rod
	78. Pin
	80. Cam
	82. Set screw
	83. Retainer
	84. Cylinder and valve assy.
	85. Seal ring
	86. Lubrication tube

Fig. IH327 — View of hydraulic lift assembly bottom side after removal.

Use care when removing the torsion bar as any dents, chips or scratches could set up stress points which might cause torsion bar failure. Bushing case (1), bushing liner (2), "O" ring retainer (3) and "O" ring (4) renewal can now be accomplished and procedure for doing so is obvious.

Reassemble by reversing the disassembly procedure.

289. SEAT SUPPORT. The right hand seat support contains the system relief valve and an unloading valve and in most cases, these components can be removed for service without removing seat support.

To remove the unloading valve, remove snap ring (22—Fig. IH330), unloading valve seat (19) and piston (6). The seat, ball, follower and

spring can be removed from carrier for cleaning; however, unit is available only as an assembly. Relief valve (17) can be removed at any time. Relief valve adjusting plug is heavily staked in place and cannot be removed. Faulty relief valves must be renewed and valves can be identified by having a "1600" stamped on head of valve used on series 706, 2706, 756, 2756, 806, 2806, 856 and 2856 or a "2000" stamped on head of valve used on series 1206, 21206, 1256, 21256, 1456 and 21456.

Check the unloading valve piston (6) as follows: Shake piston. Piston must not rattle nor should rod be loose or easy to turn. Place unit in a soft jawed vise and compress. Piston unit must compress at least 0.045. Blow through orifices in piston to insure that they are open and clean. If piston does not meet all of these conditions, renew the complete unit.

Reassemble by reversing the disassembly procedure.

290. QUADRANT AND LEVER ASSEMBLY. The quadrant and lever assembly can be removed and disassembled after removal of tractor seat, and if desired, the right fender.

Remove lever knobs, then unbolt quadrant and slide it off levers. Use two small punches to loosen lock nuts (B—Fig. IH331) to relieve spring tension. Remove cover plate (C) and side plate (E). Move lever to place rod (F) in uppermost position, then use a pair of vise grips through side opening to hold rod in this position. Remove the retaining rings (G). Remove the two cap screws (D) and the Allen head screws located under cover (C). Control lever assemblies can now be pulled out for service. Do not mix the friction discs or alter position control linkage adjustment. If bushing or seal in quadrant support need renewal, remove support (J) from seat support.

When reassembling, note all marked splines and adjust friction disc tension as outlined in paragraph 272.

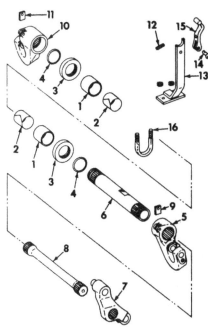

Fig. IH329—View of parts which comprise the load (draft) sensing mechanism on series 1206, 21206, 1256, 21256, 1456 and 21456.

1. Bushing case	10. Crank arm, LH
2. Bushing liner	11. Retainer
3. "O" ring retainer	12. Adjusting screw
4. "O" ring	13. Lower sensing
5. Crank arm, RH	arm
6. Crank arm shaft	14. Pin
7. Anchor bracket	15. Upper sensing
8. Torsion bar	arm
9. Retainer	16. "U" bolt

HYDRAULIC LIFT PUMP

All Models

291. The pump for the hydraulic lift system is located in the left forward end of the differential portion of tractor rear main frame. Pump is driven from a gear mounted on aft end of the pto driven shaft. See Fig. IH332. Pump can be either 12 gpm or 17 gpm capacity on all series tractors and the basic difference between the two pumps is the width of gear teeth. The 17 gpm pump is generally installed on International model tractors which are being used for industrial purposes.

292. R&R AND OVERHAUL. To remove the hydraulic pump, drain the

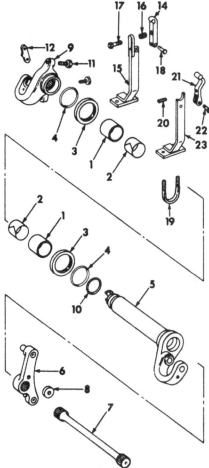

Fig. IH328—View of lower load (draft) sensing mechanism on series 706, 2706, 756, 2756, 806, 2806, 856 and 2856. Items 14 thru 18 were used on early production tractors.

1. Bushing case	14. Upper sensing arm
2. Bushing liner	15. Lower sensing
3. "O" ring retainer	arm
4. "O" ring	16. Spring
5. Crank arm, RH	17. Adjusting screw
6. Anchor bracket	18. Pin
7. Torsion bar	19. "U" bolt
8. Retainer	20. Adjusting screw
9. Crank arm, LH	21. Upper sensing arm
10. "O" ring	22. Pin
11. Stop screw	23. Lower sensing
12. Retainer	arm

Fig. IH330 — Exploded view of right seat support when tractor is equipped with load control hydraulic lift system.

1. Valve end cover
4. Plug
5. "O" ring
6. Piston
7. "O" ring
8. Seat support
9. Bushing
10. Pipe plug
11. Plug
12. Cover
13. Gasket
14. "O" ring
15. Back-up washer
16. "O" ring
17. Relief valve
18. "O" ring

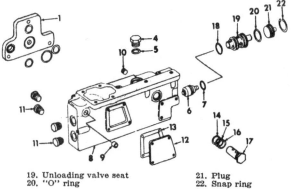

19. Unloading valve seat
20. "O" ring
21. Plug
22. Snap ring

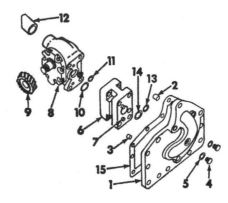

A. Quadrant
B. Lock nuts
C. Cover
D. Cap screw
E. Cover
F. Lower rod
G. Retainer rings
H. Friction discs
J. Support

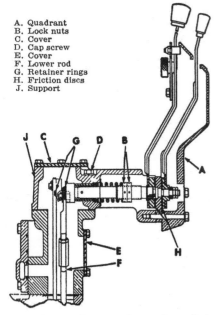

Fig. IH331 — Cross-sectional view of the quadrant and support assembly.

Fig. IH333—View of hydraulic lift pump, spacer and mounting flange.

1. Mounting flange
2. Dowel
3. Dowel
4. Plug
5. "O" ring
6. Spacer
7. Dowel
8. Pump
9. Drive gear
10. "O" ring
11. "O" ring
12. Suction tube
13. "O" ring
14. "O" ring
15. Gasket

tractor rear main frame, then unbolt and remove mounting plate, spacer and pump. Pump and spacer can now be separated from mounting plate by removing the attaching cap screws. See Fig. IH333.

293. OVERHAUL CESSNA. With pump removed from spacer, proceed as follows: Remove pump drive gear and key, then unbolt and remove covers (2 and 14—Fig. IH334). Balance of disassembly will be obvious after an examination of the unit.

Pump specifications are as follows:

12 GPM Pump

O.D. of shafts at bushings
  (min.) 0.810

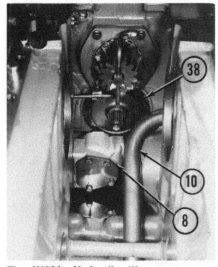

Fig. IH332—Hydraulic lift system pump (8) is mounted at left front of differential compartment. Differential has been removed for illustrative purposes.

10. Suction tube　　　　38. Drive gear

I.D. of bushings in body
  and cover (max.) 0.816
Thickness (width) of gears
  (min.) 0.572
I.D. of gear pockets (max.) 2.002
Max. allowable shaft to
  bushing clearance 0.006

17 GPM Pump

O.D. of shafts at bushings
  (min.) 0.810
I.D. of bushings in body
  and cover (max) 0.816
Thickness (width) of gears
  (min.) 0.813
I.D. of gear pockets (max.) 2.002
Max. allowable shaft to
  bushing clearance 0.006

When reassembling, use new diaphragms, gaskets, back-up washers, diaphragm seal and "O" rings. With open part of diaphragm seal (5) towards cover (2), work same into grooves of cover using a dull tool. Press protector gasket (6) and back-up gasket (7) into the relief of diaphragm seal. Install check ball (3) and spring (4) in cover, then install

diaphragm (8) inside the raised lip of the diaphragm seal and be sure bronze face of diaphragm is toward pump gears. Dip gear and shaft assemblies in oil and install them in cover. Position wear plate (15) in pump body with the bronze side toward pump gears and cut-out portion toward inlet (suction) side of pump. Install pump body over gears and shafts and install retaining cap screws. Torque cap screws to 20 ft.-lbs. for the 12 gpm pump or to 25 ft.-lbs. for the 17 gpm pump.

Check pump rotation. Pump will have a slight amount of drag but should turn evenly.

294. OVERHAUL THOMPSON. With pump removed from spacer, proceed as follows: Remove pump drive gear and key (8—Fig. IH335), then unbolt and remove pump covers (2 and 15). Bearings (7), pressure plate spring (6), "O" ring retainers (5), "O" rings (4), back-up washers (3) and oil seal (1) can now be removed from cover. Note location of bearings (7) so they can be reinstalled in the same position. Remove "O" rings (11 and 12), wear plate (16) and the pump gears and shafts (9) from pump body. Wear plate (16) is installed with reliefs toward pressure side of pump.

Pump gears and shafts, as well as the pump shaft bearings, are available only in sets. Except for suction port "O" rings (12), none of the gaskets or "O" rings are available separately.

Pump specifications are as follows:

12 GPM Pump

O.D. of shafts at bearings
  (min.) 0.812
I.D. of bearings in body
  and cover (max.) 0.816
Thickness (width) of gears
  (min.) 0.7765
I.D. of gear pockets (max.) 1.772
Max. allowable shaft to
  bearing clearance 0.004

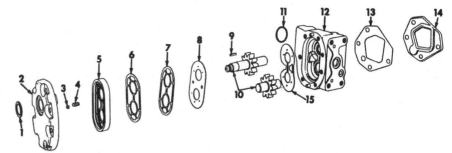

Fig. IH334—Exploded view of Cessna pump. Pump may be either 12 or 17 gpm capacity.

1. Oil seal
2. Cover
3. Ball
4. Spring
5. Diaphragm seal
6. Protector gasket
7. Back-up gasket
8. Pressure diaphragm
9. Key
10. Gears and shafts
11. "O" ring
12. Body
13. Gasket
14. Rear cover
15. Wear plate

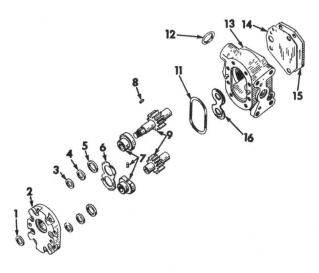

Fig. IH335 — Exploded view of Thompson pump. Pump may be either 12 or 17 gpm capacity.

1. Oil seal
2. Cover
3. Back-up washer
4. "O" ring
5. Retainer
6. Pressure plate spring
7. Bearings
8. Key
9. Gears and shafts
11. "O" ring
12. "O" ring
13. Body
14. Gasket
15. Rear cover
16. Wear plate

17 GPM Pump

O.D. of shafts at bearings
(min.) 0.812
I.D. of bearings in
body and cover (max.).......0.816
Thickness (width) of gears
(min.) 1.072
I.D. of gear pockets (max.).....1.772
Max. allowable shaft to
bearing clearance...........0.004

Lubricate all parts during assembly, use all new gaskets and seals and be sure bearings in cover are reinstalled in their original positions, if same bearings are being installed. Tighten cover to body cap screws to a torque of 20 ft.-lbs. for the 12 gpm pump, or 30 ft.-lbs. for the 17 gpm pump.

Check pump rotation. Pump will have a slight amount of drag but should turn evenly.

AUXILIARY CONTROL VALVE

All Models

295. Tractors may be equipped with either single or double control valve auxiliary hydraulic systems. Control valves are mounted on inner side of the right seat support and all valves used are similar. Therefore, only one valve will be discussed.

Hydraulic flow for the auxiliary valves is supplied by the same pump that provides fluid for the implement hitch. All of the hydraulic fluid returning to the reservoir passes through the center ports of the auxiliary valve when valve is in neutral. However, when valve is actuated, the flow through this port is stopped, the fluid directed to one of the power beyond outlets and the system immediately goes on pressure.

296. **R&R AND OVERHAUL.** To remove the auxiliary control valve, or valves, first remove seat and base as an assembly. Remove the front and rear support covers and the left seat support. Remove unions and separate tubing from top of valve. Disconnect linkage from valve spool. Remove the attaching cap screws, hold tubing up away from valve and remove valve from tractor.

When reinstalling, tighten the mounting cap screws to 20-25 ft.-lbs. torque in 5 ft.-lbs. increments. All mounting cap screws must be tightened exactly the same.

To disassemble, use Fig. IH336 as a guide. Remove end cap (1), then unscrew the actuator (10) and remove the actuator and detent assembly. Remove sleeve (15) and pull balance of parts from body. Check valve retainer (22) and poppet (20) assembly can be removed after removing snap ring (25).

NOTE: Some valves may not include the detent assembly. When disassembling these valves, sleeve (15) must be removed before removing actuator (10).

In addition, industrial valves have a circuit relief valve located directly below sleeve (15) and valve can be removed at any time.

Detent (3, 4, 5 and 6) can be disassembled after removing plug (2). Push unlatch piston (8) out of actuator (10) with a long thin punch.

NOTE: Unlatch piston (29), back-up washer (26), "O" ring (27) and plug (28) are used on series 1456 and 21456 instead of unlatch piston (8).

Inspect all parts for nicks, burrs, scoring and undue wear and renew parts as necessary. Spool (18) and body (30) are not available separately. Check detent spring (3) and centering spring (12) against the following specifications:

Detent spring
Free length-inches ....$1\frac{1}{16}$
Test load lbs.
at length-in.....23.5-28.5 @ 45/64
Centering spring
Free length-inches ....$2\frac{5}{16}$

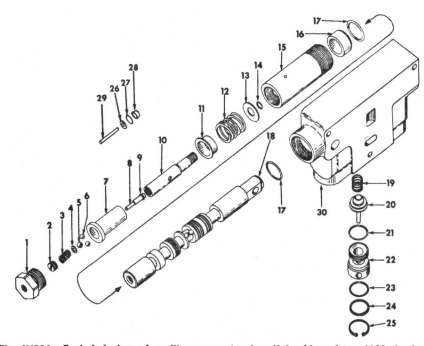

Fig. IH336—Exploded view of auxiliary control valve. Unlatching piston (29), back-up washer (26), "O" ring (27) and plug (28) are used on series 1456 and 21456.

1. Cap	8. Unlatching piston	16. "O" ring retainer	23. "O" ring
2. Adjusting plug	9. "O" ring	17. "O" ring	24. Back-up washer
3. Detent spring	10. Actuator	18. Spool	25. Snap ring
4. Washer	11. Spring retainer	19. Poppet spring	26. Back-up washer
5. Actuating ball	12. Centering spring	20. Poppet	27. "O" ring
6. Detent ball (3 used)	13. Washer	21. "O" ring	28. Plug
7. Sleeve	14. "O" ring	22. Check valve retainer	29. Unlatching piston
	15. Sleeve		30. Valve body

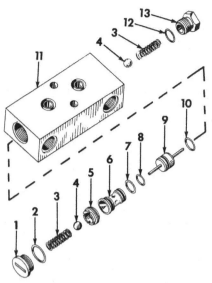

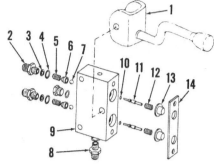

CHECK VALVE

All Models

297. A double acting check valve is used with auxiliary system rear outlet which checks the flow of fluid in both directions and precludes the possibility of equipment dropping either during transport or while parked.

Removal and disassembly of the unit will be obvious upon examination of the unit and reference to Fig. IH337.

HYDRAULIC SEAT

All Models So Equipped

298. A hydraulic controlled seat attachment is available. Fluid is supplied from a tee connection in the brake supply line. See Fig. IH338. Fluid flows to the seat control valve only when seat is being raised. The speed of raise is controlled by a 0.054-0.058 drilled orifice in the control valve. Return oil from the single action cylinder is dumped back in reservoir. Seat ride control is adjustable and is controlled by rotating the ride control valve knob. This needle valve adjusts the variable orifice opening between the seat cylinder and a nitrogen filled accumulator.

299. **CONTROL VALVE.** To disassemble the control valve, remove pivot pin and control lever (1—Fig. IH339). Unbolt and remove plate (14), then withdraw caps (13), springs (12) and pistons (11) with "O" rings (10). Unscrew connectors (2) and remove washers (3), springs (5), retainers (6) and balls (7). Clean and inspect

Fig. IH337—Exploded view of double acting check valve assembly.

1. Plug	8. "O" ring
2. "O" ring	9. Piston
3. Spring	10. "O" ring
4. Ball	11. Block
5. Nut	12. "O" ring
6. Retainer	13. Plug
7. "O" ring	

Test load lbs.

at length-in. .....26.5-33.5 @ 1-7/64

Use all new "O" rings and reassemble by reversing the disassembly procedure. Detent unlatching pressure is adjusted by plug (2). Unit must unlatch at not less than 1550 psi nor more than 1750 psi or series 1456 and 21456, or not less than 1000 nor more than 1250 psi on all other series. The circuit relief valve on industrial valves is a cartridge type with the pressure setting stamped on end of body. Faulty relief valves are corrected by renewing the complete unit.

Fig. IH339—Exploded view of hydraulic seat control valve.

1. Control lever	8. Connector
2. Connector	9. Body
3. Washer	10. "O" ring
4. "O" ring	11. Piston
5. Spring	12. Spring
6. Retainer	13. Cap
7. Ball	14. Plate

all parts for excessive wear or other damage.

Renew all "O" rings and reassemble by reversing the disassembly procedure.

300. **CYLINDER.** Disassembly of the single action seat cylinder is obvious after an examination of the unit and reference to Fig. IH340. Clean and inspect all parts for excessive wear or other damage.

When reassembling, use new "O" rings (3, 6, 8 and 10) and new wiper ring (4).

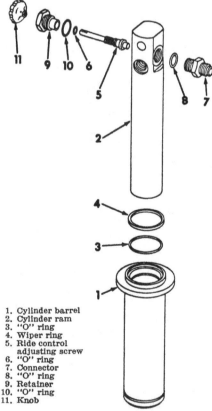

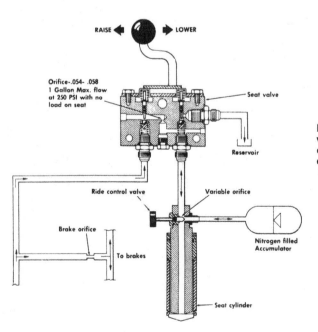

Fig. IH338 — Schematic view of hydraulic circuit, control valve and cylinder of the hydraulically controlled seat attachment.

Fig. IH340—Exploded view of single action seat cylinder.

1. Cylinder barrel
2. Cylinder ram
3. "O" ring
4. Wiper ring
5. Ride control adjusting screw
6. "O" ring
7. Connector
8. "O" ring
9. Retainer
10. "O" ring
11. Knob